Dev Chen
Berkeley '87

SIGNAL PROCESSOR CHIPS

D1477580

SIGNAL PROCESSOR CHIPS

David Quarmby
editor

Prentice-Hall, Inc., Englewood Cliffs, New Jersey 07632

Library of Congress Cataloging in Publication Data

Main entry under title:

Signal processor chips.

"A Spectrum Book."

Includes bibliographies and index.

1. Signal processing—Digital techniques.
2. Integrated circuits. I. Quarmby, David J.

TK5102.S543 1985 621.38'043 84-18287
ISBN 0-13-809450-0
ISBN 0-13-809443-8 (pbk.)

© 1985 by Prentice-Hall, Inc., Englewood Cliffs, New Jersey 07632; Granada Publishing Limited, London, England; and David J. Quarmby. All rights reserved. No part of this book may be reproduced in any form or by any means without permission in writing from the publisher.
A Spectrum Book. Printed in the United States of America.

Previously published by Granada Publishing Limited, London, England. Copyright © D.J. Quarmby 1984.

This book is available at a special discount when ordered in bulk quantities. Contact Prentice-Hall, Inc., General Publishing Division, Special Sales, Englewood Cliffs, N.J. 07632.

10 9 8 7 6 5 4 3 2 1

ISBN 0-13-809450-0
ISBN 0-13-809443-8 {PBK.}

Prentice-Hall International, Inc., *London*
Prentice-Hall of Australia Pty. Limited, *Sydney*
Prentice-Hall Canada Inc., *Toronto*
Prentice-Hall of India Private Limited, *New Delhi*
Prentice-Hall of Japan, Inc., *Tokyo*
Prentice-Hall of Southeast Asia Pte. Ltd., *Singapore*
Whitehall Books Limited, *Wellington, New Zealand*
Editora Prentice-Hall do Brasil Ltda., *Rio de Janeiro*
Prentice-Hall Hispanoamericana, S.A., *Mexico*

CONTENTS

PREFACE

The tide of digital electronics continues to engulf more and more of electrical engineering. It moves in surges, and a surge is now taking place in the implementation of signal processing electronics. Every industrialised nation seems to have its own national plan, and there are a few international ones as well, to move into the 'next generation' of computers. This envisages a quantum jump in the 'intelligence' of the machine, and the convenience of its interface with man. These objectives are fuelling great activity in the production and application of integrated circuits for digital signal processing, the subject of this book.

We centre on three programmable signal processors, produced and sold by three of the world's largest chip manufacturers: Intel and Texas Instruments of the USA, and the Nippon Electric Company of Japan. Other devices have been produced by other companies, but many are for internal use only and have made little impact on the world at large. The authors who have contributed the central chapters on these three devices, the Intel 2920, TMS 320 and NEC 7720 are international experts in the application of the devices, and each holds a senior position with the manufacturer of the chip concerned. They are busy men and I am most grateful that they have spared the time and effort to contribute to the book.

We expect our readers to be engineers, professionals or aspiring professionals, graduates or students in the later years of their courses. Our approach is a very practical one. The style is more akin to the literature of the chip manufacturer than to an academic text. Detailed mathematics has been totally avoided. It is amply covered elsewhere. Algorithms have been described at a flowchart level, with references to mathematical texts should these be needed. The chapter on algorithms was written with the intention of covering only those applications used as examples in later chapters. In fact, it turned out that it had to be fairly exhaustive, and certainly covers the commonly used algorithms.

There is currently a rather small fraternity using these devices. It is growing rapidly, and I hope that our efforts will spread the enthusiasm of the authors and swell the number of users.

David J. Quarmby, B.Sc., Ph.D.

SIGNAL PROCESSOR CHIPS

CHAPTER 1
BACKGROUND AND APPLICATIONS

D. J. Quarmby, Loughborough University

1.1 A New Type of Chip

The dominant development in electronics during the 1970s was the microprocessor chip. It captured the imagination of the whole engineering world, and there can be few practising engineers who are unaware of its potential. The word 'microprocessor' has become a buzzword, now entering into everyday use.

To the computer engineer, miniaturisation of the processor was just one of a large number of important steps which have led to the miniaturisation of the computer as a whole. There has been no huge breakthrough of principle. The ideas embodied in the latest chips are exactly those which have been used in the huge machines of the past. What is new is the range of applications now opened up to thc influence of digital computing.

Signal processing is an area where the application of microelectronics has been slow to make an impact. The computer has been used to process analog signals for some time now, but the high cost of high speed digital arithmetic units has in the past restricted this use to either slowly changing signals, off-line processing of signals, or applications where high expense can be justified.

The picture has suddenly changed. A new breed of chip has emerged, and it is the purpose of this book to help engineers to appreciate this change and to adjust their thinking to include these devices.

Chip makers exist to make profits by selling their products in large numbers. They must see areas of application where their devices will be used before they embark on the long and expensive design road. One such application area which has been thrown wide open by these new devices is that of speech processing. It is this market which has made it worthwhile to depart from the microprocessor architecture of the 1970s, and to try out some new ones. Speech synthesis has been the first application to reach mass markets. The chips used here have by and large been specially designed for that purpose, and cannot be programmed to perform a variety of functions. Texas Instruments lead the way both in devices and applications, with their LPC synthesis chips (TMS 5100, 5200 and 5220) and the 'speak and spell' educational toy. Video games, home computers, cars, and many other machines now incorporate similar chips. Speech quality remains rather indifferent, and in the effort to improve it, a number of more complicated experimental speech synthesisers have been produced (ref. 1.1). This is where the new programmable signal processors come into their own. The designs are not quite ready to go directly to a special chip, yet are sufficiently well tried to be programmable at the very low level used in the chips which we are describing. The majority of the current applications of speech synthesis use a stored set of control parameters to generate a limited number of phrases. They act like a solid-state recording, with fast random access to each phrase. This will be only a temporary phase, though there are varying opinions on its length. Already there are cheap systems available which will produce speech directly from stored text. Commercially available units produce speech which is only just intelligible, but far better systems are in the research laboratories, and the signal processor chips have their role to play in bringing these to market.

Speech synthesis is coming of age. Speech analysis and recognition are still in infancy. Signal processor chips are finding a major role in this area. An advantage of the LPC synthesiser is that the corresponding analyser can be readily defined. Already analysers have been implemented in these programmable chips (refs. 1.2, 1.3). The analyser/synthesiser combination is the vocoder, used to reduce the speech signal to a low data rate for transmission. The use of very low bit rate vocoders has been rather restricted, but as more and more telephony becomes digital, the demand for cheap vocoders, transmitting speech of reasonable quality, will increase. Speech recognition has so far reached few applications. The machines are expensive, they require a period of 'training' to adapt to a new vocabulary and new speaker, and only a few can deal with continuous speech. The signal processor chips will immediately make an impact on the size and expense of such systems. The operations which they perform well are those of filtering, spectral analysis and correlation – all major components of speech recognition. The rewards open to the manufacturers of speech recognition systems are high, and we will see rapid progress in this area.

Speech processing is not the only application area for these devices. Another major area is in digital data transmission – in the design of the modem. The detection and correction of errors in transmission is a topic which has been well developed theoretically (ref. 1.4). The algorithms deal mainly with digital information, at data sampling rates similar to those of the speech signal. Our chips are ideal for experimental systems and can later be mask programmed for production runs.

Another application area where the signal bandwidth is similar to that of speech is low frequency sonar signal processing. As the signal bandwidth increases, the sophistication of processing which can be done will naturally decrease, but in my own laboratory, students have carried out simple graphics generation using a signal processor chip to directly produce the video signal.

The impact of the programmable signal processor chip in these application areas will be felt over the next few years. The impact may well be such that the 'signal processor' will enter the jargon as the microprocessor has. Yet once again, there is no major revolution in principle. Integrated circuit technology has reached a stage where a high speed multiplier can be included with all the rest of a computer on one chip. There have been adjustments to the usual computer architecture in order to make good use of the multiplier and other computing resources. It has been argued (ref. 1.5) that the signal processor chips are just pathfinders in the steady evolution of the general-purpose computer. Certainly we will soon see the fast multiplier incorporated in general-purpose microprocessor chips. Certainly this will mean that this new generation of microprocessors will then be capable of handling speech signals in a similar way to the current signal processors. It may well be, however, that the specialist signal processor computer has moved on to tackle signals of higher bandwidth, and that the architectures will diverge rather than converge.

For the time being we confine ourselves to a look at the currently available devices, where they have their origins, what they can do and how they do it.

1.2 Computers in the Analog World

The digital signal processing engineer lives in a world of precision. The signals he processes unfortunately arise from transducers which generate an analog voltage. The conversion from the real world to the utopian digital world must be accomplished as painlessly as possible.

Chip manufacturers are performing wonders by providing components which require little knowledge of analog circuit techniques.

The first requirement for analog circuitry concerns the problem of aliasing. It is fairly obvious that if one is dealing with a rapidly changing high frequency signal, one will have to take samples of that signal at a high rate. It is clearly going to be no use sampling a speech signal at 100 Hz when the frequencies of interest stretch up to 4 or 5 kHz. To deal with such a signal one must sample at more than twice the highest frequency of interest, in this case at 10 kHz, the presence of over 2 samples per cycle being enough to allow the sinusoid to be re-created. What is less obvious is that this in itself is not enough. If

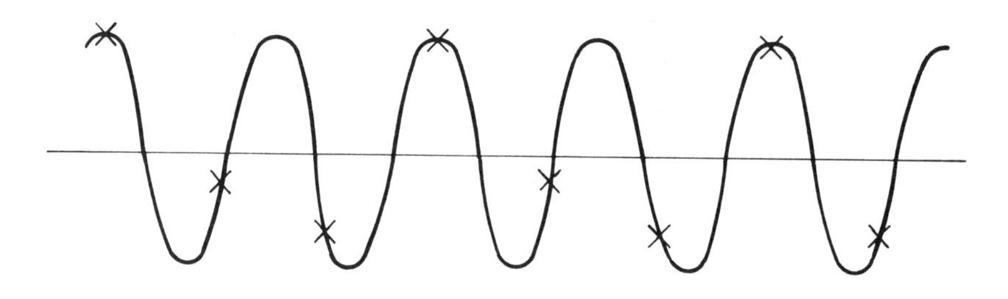

a) Original signal ($6\frac{2}{3}$ KHz) sampled at 10 KHz

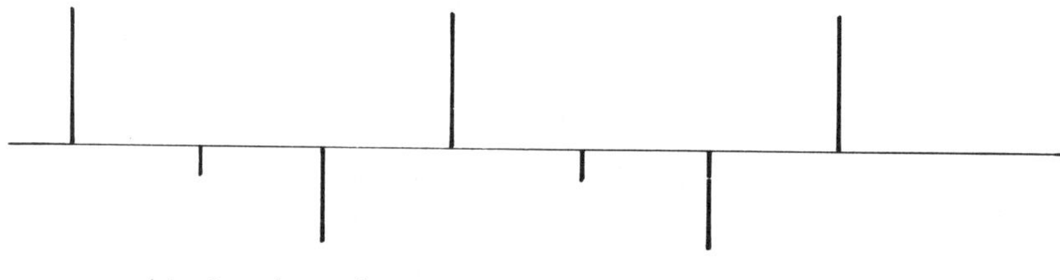

b) Sample values

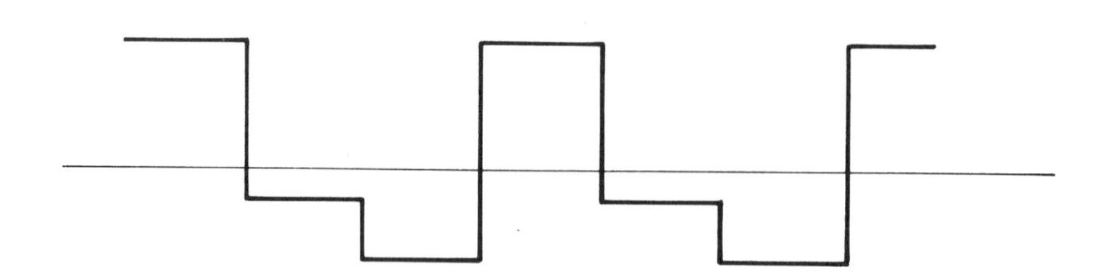

c) Signal recreated from sample values, fundamental frequency now $3\frac{1}{3}$ KHz.

Figure 1.1 Aliasing

there are any frequency components present which are higher than half the sampling frequency, they will be interpreted as if they were lower frequencies, as fig. 1.1 shows. Here, a sinusoid of $6\frac{2}{3}$ kHz is being sampled at 10 kHz. It can be seen that the resulting samples could equally well have come from a sinusoid of $3\frac{1}{3}$ kHz. The ambiguity cannot be allowed, and the digital processing would have to assume that it was the lower frequency which was present, as this is within the region of interest.

It is possible that in some applications this problem will not arise. The transducer itself will be limited in its response rate, and it may be possible to sample at such a high rate that aliasing can be guaranteed not to be a problem. If aliasing is possible, it must be eliminated by first passing the signal through a low pass filter. Ideally this will be flat up to the cut-off point, and have zero response thereafter. This is not possible in practice, and the system designer must decide on two limiting frequencies. The lower one is the limit of frequencies which must be preserved with some specified tolerable deviation from the ideal constant gain. The upper limit is half the sample rate, and above this figure all frequencies must be attenuated so that they can in practice be neglected. The wider the region between these two limits, the easier it is to design the analog filter. For the lazy digital engineer the manufacturers provide easy-to-use filters of high specification. A particularly appealing type for the digital designer is the switched-capacitor filter, in which the cut-off frequency can be controlled by varying a (much higher) clocking frequency to the device. Typical examples of these are the Intel 2912 and Motorola MC145414.

Low pass filtering is also required when a signal is being created by a digital system. A speech synthesiser is a typical example of this situation. The signal produced by a D/A converter is stepped, each step being held for one sample period. Once again it may be that the stepping is so fast that the output transducer (in this case a loudspeaker) will itself smooth out the steps. If not, a low pass filter of similar specification to the anti-aliasing filter will be needed. This is usually called an 'interpolation' filter for obvious reasons.

The subject of A/D and D/A converters is best covered by the literature of the specialist manufacturers who produce these devices. The devices are becoming smaller and easier to use, with many requiring very few external components. An important question, when using an A/D concerns the need for a sample/hold device. For very low frequencies it can often be avoided. A 'tracking' type of converter can be used, in which the digital value is stepped up or down by one unit only between successive comparisons of the voltage it represents (generated using a D/A) and the incoming voltage (fig. 1.2 (a)). A high resolution device will clearly take a substantial time to count from one extreme to the other, severely limiting the slew rate of the device. At the other end of the speed scale, in the nanosecond range, the 'flash'

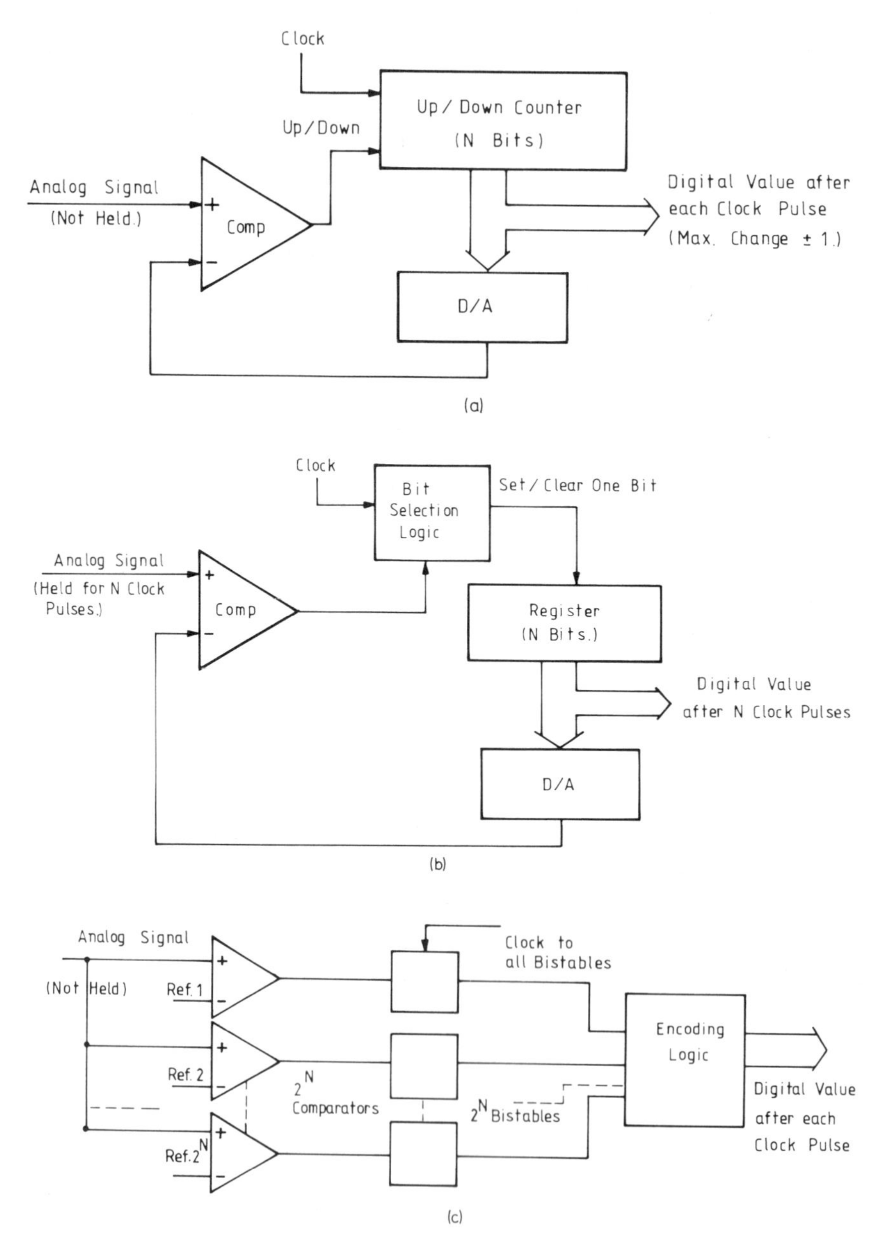

Figure 1.2 (a) Tracking A/D; (b) Successive approximation A/D; (c) 'Flash' A/D

type of converter is equally easy to use, and needs no sample/hold. These converters use a large number of comparators, the incoming signal being compared with, perhaps 256 preset levels – all in parallel. The resulting 256 bit pattern can then be suitably encoded (fig. 1.2 (c)) to an 8-bit value. Between the two extremes of speed, in the region of most interest to us, the most frequently used device type is the successive approximation converter (fig. 1.2 (b)). This does require a sample and hold, as the conversion process sets each bit in turn, starting with the most significant one, at each stage comparing the partial result with the incoming (held) signal via a D/A converter. Without the holding of the signal, catastrophic errors would occur at those voltages where the more significant bits are changing over.

The reason for using a successive approximation device rather than the 'flash' device is that higher resolution can be obtained. The flash devices are produced with up to 9-bit accuracy, and successive approximation devices range up to 18 bits.

It is worth considering that the usual reason for a large number of bits is to accommodate a wide dynamic range of signals. Consider, for example, the echo signal amplitude produced by a piezo-electric sonar transducer (fig. 1.3). This can vary from a few volts down to microvolts, a factor of around 10^6, needing 20 bits of dynamic range. At any particular depth, however, the accuracy needed in a signal

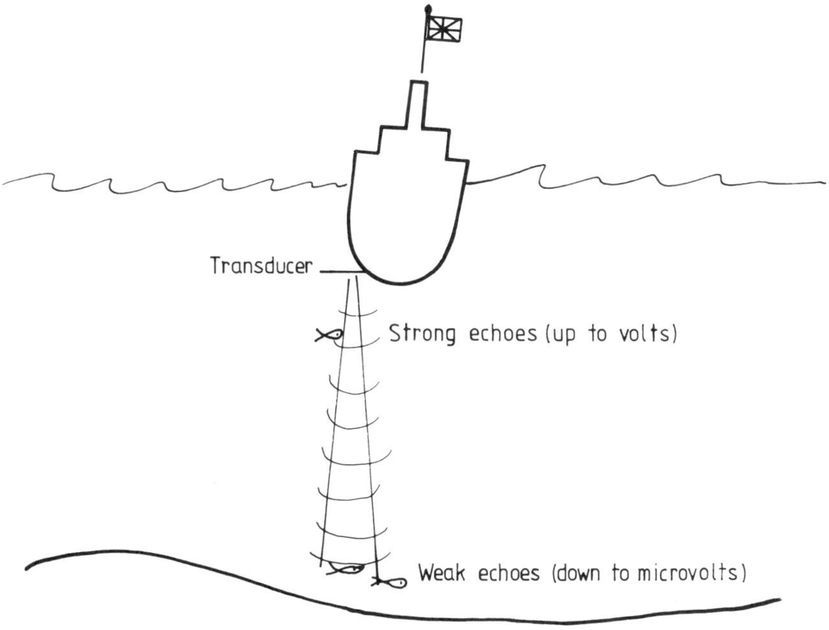

Figure 1.3 A sonar system

measurement is only perhaps 1 part in 1000, needing only 10 bits to represent it. In the sonar case, and in many other situations, the answer is to include an amplifier prior to A/D conversion. A particularly useful type of amplifier is one which can be controlled digitally, and whose gains alter in steps of powers of two. A nice example is the MN2020 produced by Micro Networks Corporation, which has three control lines which are used to provide gains between 1 and 128. These 3 bits are therefore used to add 7 bits to the dynamic range of the signals which can be converted. A 13-bit A/D would then be needed to cover the 20-bit dynamic range of our sonar system. The resulting representation is, of course, of the floating point type, with a 13-bit mantissa and a 3-bit exponent. We return to the question of number representation in chapter 2.

There is an inevitable need for analog electronic components in signal processing systems. At the very highest frequencies, digital methods have yet to make any impact. There has, however, been a major change in the way analog parts of a mainly digital system are designed, with the main building blocks required coming in easy-to-use packages. Circuit design at the low level is increasingly the province of a few people working for the chip manufacturers.

1.3 The Multiplier

No engineer who has worked with digital systems needs this book to justify their use whenever possible. In my experience, properly designed and constructed digital circuits work first time, whereas anything other than the simplest analog circuits require adjustments. Signal processing has, however, clung to its analog past for longer than other fields where the computer found immediate application. The reason was the lack of a sufficiently fast and cheap multiplier. The operation of multiplication is central to signal processing. Consider for example, the components used in filtering, involved in almost every signal processing system. In the very simplest of passive filtering components we see the multiplication operation in the form of Ohm's Law. Moving into active circuits, the amplifier working linearly gives us the multiplication effect. The idea of replacing an operational amplifier, or worse still, a resistor, by a huge array of switching logic, is one which makes even the committed digital designer think twice about his approach.

Theoretical advance was also needed. The analysis of linear systems using a continuous time approach has been very well developed, stretching back to the work of Fourier and Laplace. The adaptation of these methods to sampled-data systems is comparatively recent, with significant advances being made in the 1960s.

The theoretical, and most experimental, work has been done using general-purpose computers in simulation studies, free from real-time constraints. There are many requirements for signal processing on data which is first stored in analog form, on a tape recorder, converted and stored in digital form, and subsequently processed at leisure. The fact that the multiply operation may take a few milliseconds, and that it may take hours to analyse a few seconds of speech data, is outweighed by the conveniences provided by a general-purpose computer: high precision floating point arithmetic, high level languages to allow easy specification of algorithms, and most of the hardware hidden from the programmer. Even with the power of the present generation of signal

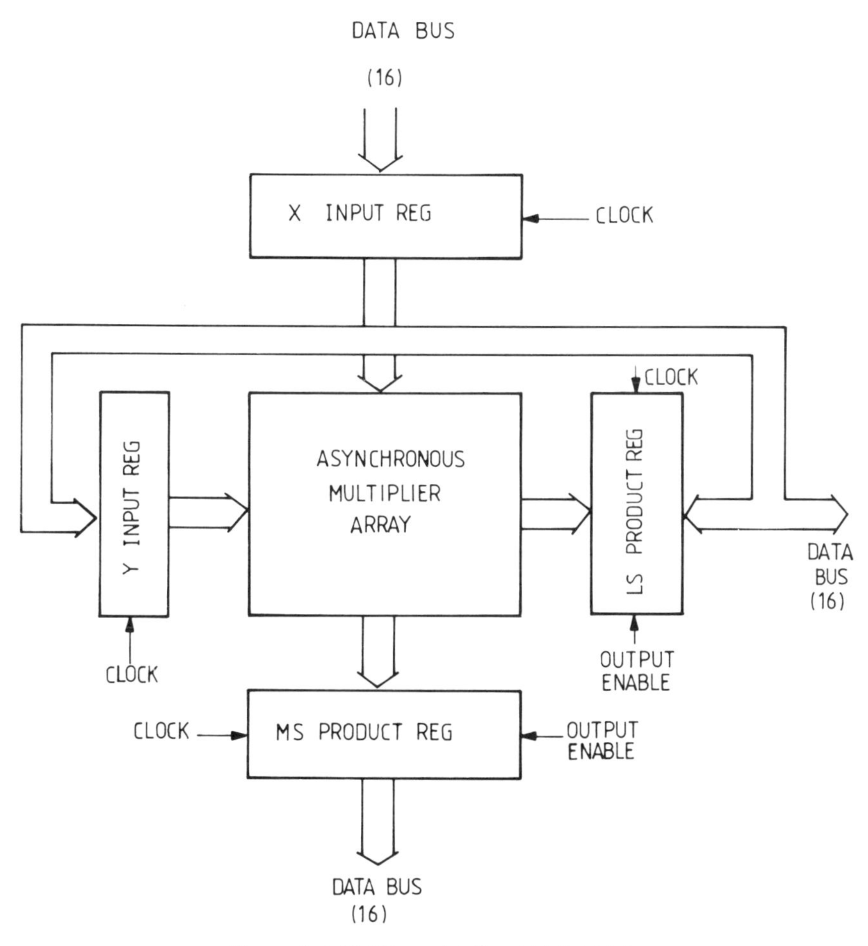

Figure 1.4 TRW multiplier MPY 16AJ

processors, this is still the best way to test new algorithms. We are concerned here with the methods of implementing these after considerable work has gone on in the simulation environment.

An important step forward in allowing digital circuitry to functionally match its analog counterpart was the building of multipliers on single chips. The TRW company lead the field with devices performing integer multiplications of 8-bit, 12-bit 16 or 24-bit numbers in a small fraction of a microsecond. Fig. 1.4 shows a block schematic of an early single chip multiplicr, the TRW MPY 16AJ. It can be seen immediately that it needs a large number of legs, 64 in fact. Its high speed circuitry also consumes considerable power, 7.5 watts. It therefore needs very good heat sinking arrangements. All this is clumsy by the standards of digital circuit engineers, but small inconveniences were happily tolerated when the alternative was a large board of components.

The introduction of the TRW multiplier range coincided with the growth in the use of the microprocessor chip. The general-purpose microprocessor is, however, a slow machine, and to use it to feed a multiplier which can work in, say, 100 nanoseconds, is like using a teaspoon to stoke a steam engine. The shovel required comes in different forms, but a popular one is the 'bit-slice' range of components. The next section outlines a typical system of this type.

1.4 A 'Bit-slice' Signal Processor

The bit-slice devices are primarily used in the design of general-purpose minicomputers. The devices are sufficiently flexible, however, for special-purpose signal processing machines to be produced using them. A number of such machines have been built in various research laboratories. To illustrate the approach we consider here the 'SPRING' system which was designed and built at Loughborough University for a sonar application, by J. P. Smith and J. W. Goodge. The acronyms were rather far-fetched but suffice it to say that the components of the system are the MARCH hardware, the APRIL monitor program, running in a small Intel 8085 host, and the MAI assembler program, set up by A. Johansson on a PDP 10 – the Swedish spelling is essential.

Fig. 1.5 shows a block diagram of the MARCH hardware. The program to be executed is held quite separately from the data on which it operates in the CONTROL MEMORY. This is a common feature of most signal processing machines, including the single chips, and is referred to as 'Harvard' architecture. In this case, the instruc-

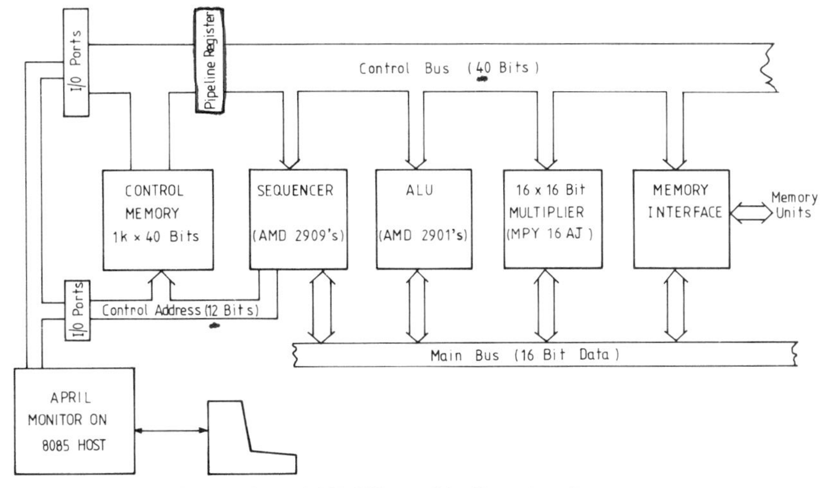

Figure 1.5 MARCH – a bit-slice signal processor

tions are 40 bits wide, and a 1K × 40-bit RAM is provided to hold them. This is high speed (HMOS) RAM. Instructions are executed at a constant clock rate of 4 MHz. Once again, a relatively wide instruction word, and a constant high instruction rate are typical of signal processing chips. Sequencing through the instructions is performed by the SEQUENCER, a subsystem which generates a 12-bit address. It is within the sequencer that we first see the use of the 'bit-slice' components. The AMD 2909, which is a complete 4-bit sequencer, was used in this design. The device allows addresses to be generated in sequence, or to be loaded from an external source (a jump) with the option of saving the current address on a built-in stack (for a subroutine call and subsequent return). A 4-bit address gives little scope for programming, but by providing appropriate links, any number of these bit-slice control sequencers can be placed side-by-side to generate any required address length. The 4-bit slice has been a popular choice, but others included 2-bit and 8-bit slices. Larger address lengths lose the flexibility of the 'slice' approach, and become complete sequence controllers in themselves.

The 40-bit output of the control memory provides the control to the rest of the system, including control to the sequencer. The action of these 40 bits on the hardware is very direct, and requires very little decoding logic. This is a strong contributor to speed, but it originated, not for speed, but for neatness and flexibility in the design of general-purpose computers, and it is necessary here to side-track, and mention the use of this type of control logic in a general-purpose computer. The idea is, in computer terms, an old one, having been put forward by

Wilkes (ref. 1.6) in 1951. The machine code of a general-purpose machine must, for reasons of economy, be efficiently coded in memory. The task of decoding it within a processor can involve a great deal of random logic. Wilkes proposed the use of an 'inner' program, within the processor, to perform this decoding task. This inner code was called a 'microprogram', and the name has remained ever since. The machine code generates a start address for a small section of microcode. The microcode instructions can afford to be much less efficient, as there are few of them, and act directly on the hardware with little or no decoding logic. Other machines have descended to further inner levels – 'nanocode' etc.

. . . And these have smaller fleas to bite 'em
And so proceed 'ad infinitum' – Swift

The instruction coding of the signal processor is usually of the 'microcode' type, and the bit-slice sequencer would normally sequence microcode. So, although there is only one level of programming in the signal processor, it is often referred to as a microcoding level.

The MARCH hardware (fig. 1.5) latches the 40-bit microprogram word in a 'pipeline register', and it is this register which actually controls system operations. Pipelining is used in many situations to speed up operations by performing just a part of an operation, then passing on the result to the next stage, etc. Henry Ford used the same idea much earlier. Pipelining, and the whole subject of bit-slice design, is thoroughly covered by Mick and Brick (ref. 1.7).

The units controlled are an Arithmetic and Logical Unit (ALU), and up to 15 'modules' which provide all other functions needed, and are each controlled in a similar manner. The ALU is again built from bit-slice components. The one used here is the AMD 2901. Once again the slice of the ALU is 4 bits wide, and contains all the logic to perform high speed addition, subtraction, logic operations and shifting. Four devices were used in this design to provide a 16-bit data word length.

Access to modules is at the 250 ns instruction rate, and so it is quite appropriate to include as a module a high speed multiplier such as the TRW devices. Other modules which have been built include a front panel, an associated 8085 microcomputer system, a high speed D/A converter for fast waveform generation, and most cruxial and complex, a memory interface. This contains two pairs of memory address registers, with appropriate logic to increment, decrement and load them. A number of hardware components are duplicated in the system to provide a very fast switch from a 'normal' state to a 'special' state in response to an interrupt – a property needed in the particular application. Fig. 1.6 shows the formats adopted for the 40-bit instructions. The first 3 bits are decoded in the pipeline register, and they indicate which of up to 8 (6 are defined) formats is to be used. At least 2 devices are controlled in each format. For instance, the first format is used to simultaneously control 3 different modules. A 12-bit field is associated with each module, 4-bits providing a module address and 8-bits providing control information. The modules themselves examine, in parallel, all three of these fields, and, finding their own address in any one of the 4-bit subfields, will pick up control information from the corresponding 8-bits. The tailoring of the instructions to suit the hardware can be seen in that module control will always occur in one or more of these three 12-bit positions, with ALU control coming in the first 24-bits (or 16-bits for partial control) and the sequencer being controlled by the last 21-bits (13-bits for partial control). Full ALU and full sequencer control cannot occur simultaneously.

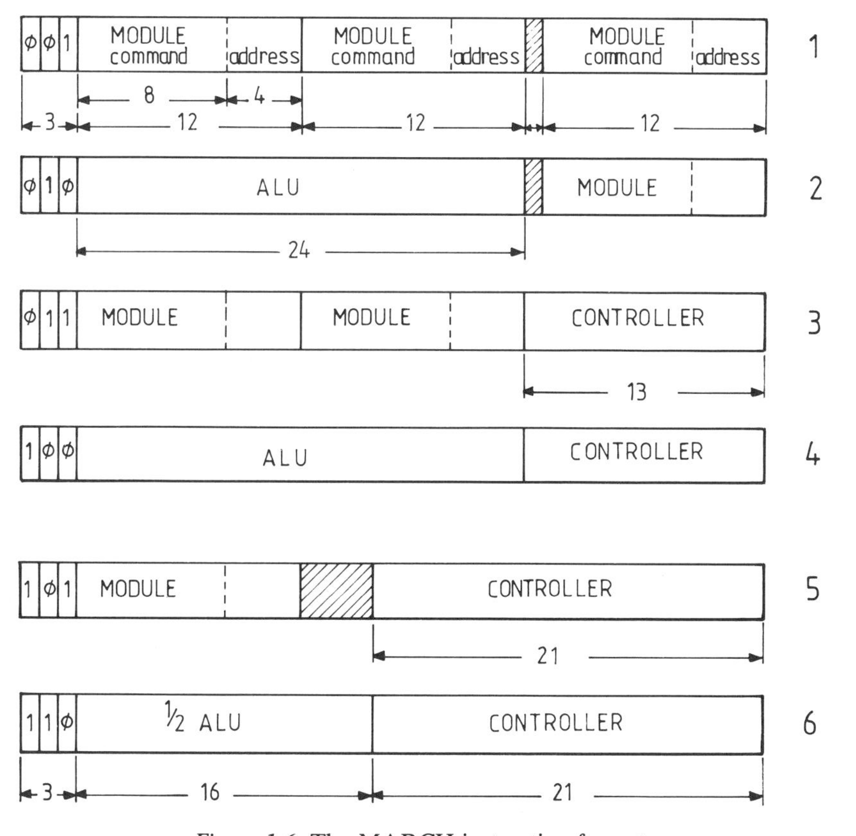

Figure 1.6 The MARCH instruction formats

Simple program loading can be effected using the 8085 system attached to the signal processor.

A typical monitor command might be:

$3 M1–F3 MF–97 C–1234

Numerical values are in hexadecimal, and the command means:

Format type 3
Module number 1, command byte F3
Module number F, command byte 97
Controller (short format), command 1234

This allows programming and program alteration at machine code level. The monitor program was in fact written in FORTH, and is called APRIL.

An assembly language was also defined, and the assembler itself was generated using a piece of software called MICGEN – written at the Royal Institute of Technology, Stockholm (ref. 1.8). This is a very powerful aid to the designer of any micro-programmed system, in that it allows the definition and simulation of a complete micro-programmed system. The definition is in terms of components which may be existing ones, or ones which the designer would like to create. The assembler–generator is just one small part of this package, which is written in the SIMULA language, initially on a PDP–10.

The SPRING system was used as an experimental sonar display unit, and incorporated special line buffers for signal entry from the sonar system and exit to a TV monitor. The sonar data uses a 10 kHz line rate and 2 or 4 Hz frame rate. The data is held in the main memory module, and read-out under program control keeps up with real-time TV video rates, but at reduced horizontal resolution, with interrupts to the signal processor coming at TV line rate.

1.5 Signal Processor Chips

The bit-slice systems are straightforward in principle, but their engineering is demanding – expensive and time consuming. The SPRING project, for instance, extended over three or four years. The hardware is relatively bulky, and coupled with the high speed logic involved, lead to a number of construction problems which took longer than expected to solve.

The manufacturers of the signal processor chips have taken away the majority of the difficulties by incorporating within a single chip much of the logic contained in a bit-slice processor. Chapters 3, 4 and 5 cover three of the chips in detail, but here it is worthwhile introducing the three and comparing them with the bit-slice approach.

The Intel 2920 was the first to be introduced, in 1979. It has a control memory of 192 × 24 bits, a very simple sequencer, and an ALU working on 25-bit data words internally (to 28-bit accuracy). Fast multiplication is assisted by a barrel-shifter. It remains the most self-contained device, having A/D and D/A convertors on the chip. The separation of program storage and data storage (40 × 25-bits) is total, the program residing in EPROM.

The NEC 7720 became available during 1981, and it too keeps program and data memory completely separate, again using EPROM for the instructions. EPROM is also used for a fixed data store. The wide instruction word (23 bits) is again apparent, and the sequencer is a sophisticated one, of similar capability to the bit-slice sequencers. The ALU handles 16-bit data, and a hardware multiplier, similar to the MPY 16AJ is built in.

The latest of the devices we consider is the TMS 320 from Texas Instruments. The separation of program and data stores is not total, but for the fastest access to storage it is used with the two separated. It has a narrower, 16-bit, instruction word although some of the most important operations involve various hardware units operating in parallel. Its ALU operates on 32 bits, and as with the NEC 7720, a sophisticated sequencer and a hardware multiplier are incorporated.

The different devices are each finding their individual roles in the new markets which they have collectively opened.

1.6 Programming Techniques

At present there is little alternative for most users than to generate code for these signal processors in their assembly language. Higher level approaches are mentioned in subsequent chapters, but are not widely available. This comes as no surprise to those familiar with new developments in computing. At this stage we must simply be grateful to have escaped from re-building the hardware whenever a change has to be made. Most users of signal processors will actually want their own variation on a relatively small number of themes – filters, spectral analysis, pattern matching, etc. We hope that the example programs in later chapters will provide a source for some of these basic requirements.

A few general programming principles are worth emphasising:

(a) Test algorithms before using a signal processor chip. The difficulty of writing low level code makes it an undesirable tool for trying out new algorithms. Experimenting on the signal processor should be confined to changing parameters and observing the result, where the benefit of immediate feedback is best used.

(b) The use of symbols to represent fixed numbers is always important. The symbols might represent:
   (i) literal values (constants)
   (ii) the address of an instruction
   or (iii) the address of a data variable.

   With the physical separation of instruction and data storage, these fixed numbers can often be of quite different word lengths. Whether or not this is so, a wise programmer will carefully keep track of the type of constant which each symbol represents.

(c) Finally a comment on comments. The lower the language level, the more they are needed, and the language level here is very low.

References

1.1 Quarmby, D.J. and Holmes, J.N. (1984) 'Implementation of a Parallel-formant Speech Synthesiser using a Single-chip Programmable Signal Processor', special issue of *IEE Proceedings (F)* on 'The Design and Application of Digital Signal Processors'.
1.2 Feldman, J. (1982) 'A Compact Digital Channel Vocoder using Commercial Devices' *Proceedings of ICASSP '82, Paris,* 1960–1963.
1.3 Feldman, J., Hofstetter, E.M. and Malpass, M.L. (1983) 'A Compact, Flexible LPC Vocoder Based on a Commercial Signal Processing Microcomputer' *IEEE Trans ASSP-31,* No. 1, 252–257.
1.4 Clark, A.P. (1976) *Principle of Digital Data Transmission* Pentech Press.
1.5 Morris, L.R. *Digital Signal Processor Software* dsps inc.
1.6 Wilkes, M.V. (1951) 'The Best Way to Design an Automatic Calculation Machine', *Manchester University Computer Conference,* 1951.
1.7 Mick, J. and Brick, J. (1980) *Bit-slice Microprocessor Design* New York: McGraw-Hill.

1.8 Persson, M. (1979) 'Design of software tools for microprogrammable microprocessors', *TRITA-NA-7903*, Dept of Numerical Analysis and Computing Science, Royal Institute of Technology, Stockholm, Sweden.

CHAPTER 2
ALGORITHMS

D. J. Quarmby, Loughborough University

2.1 Variable Representation

All three of the signal processor chips to be discussed in detail use a fixed point representation, although the word length of the arithmetic unit is different in all three.

As discussed in section 1.2, the majority of signals span a wide dynamic range, and the floating point representation is a much more natural representation. Fixed point machines can of course be used to perform floating point arithmetic, and so it is worth reviewing the operations involved.

A floating point variable (V) is represented by two independently stored integers, usually both signed, the mantissa (M) and exponent (E) such that

$$V = M \cdot B^E$$

where B is the base to be used throughout. We will consider only the case where $B = 2$, the most commonly used base

$$V = M \cdot 2^E$$

Once E becomes fixed, at whatever value, the system being used is a fixed point system, in which the variable is represented by the integer M and an implied scaling factor 2^E.

Consider the situation where integers are stored as 16-bit (15-bit + sign) numbers, and the signal is a 14-bit (13-bit + sign) integer read from an A/D converter (fig. 2.1(a)). A common requirement (in correlation, for example) is to multiply two signal values together. The product would be a 31-bit number (2 × 15-bit + sign), of which only 27 bits (2 × 13 + sign) are of

S = sign bit
X_j = bit j of signal value
P_j = bit j of a product value

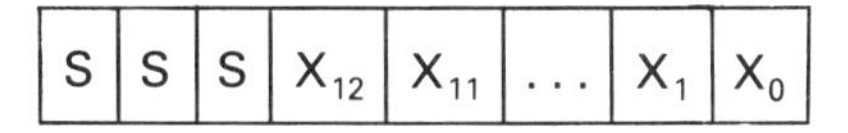

For small signals some of X_{12}, X_{11}, etc. = S
(a). 14-bit signal value in 16-bit word.

S	S	S	S	S	S	P_{25}	P_{24}	. . .	P_1	P_0

For small signals, some of P_{25}, P_{24}, etc. = S
(b). Product of 2 signals in 32-bit word.

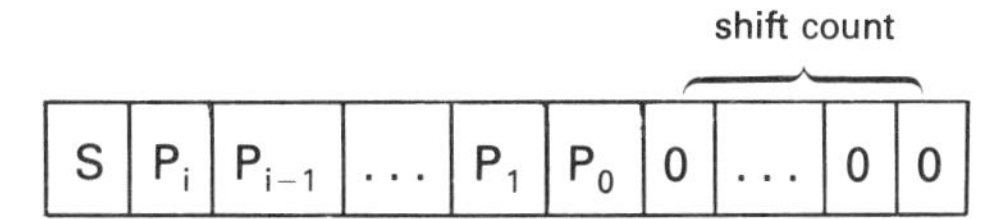

P_i is leftmost bit $\neq$ S.
Shift count stored as exponent.
Product can be truncated to 16-bits with minimum loss of accuracy.
(c). Normalised product in 32-bit word.

Figure 2.1

interest to us (fig. 2.1(b)). The dilemma arises if we have no choice but to store the product in a 16-bit variable. We must arrange things so that we never suffer an overflow in calculation, or things will go catastrophically wrong, yet if we do this by taking the sixteen most significant bits of interest, we totally lose our signal when it has a low value. With the problem as posed here, we have no choice but to normalise the product by shifting it (flowchart 2.1) one step at a time, to the left, until the first digit of the number occupies the most significant position (fig. 2.1(c)). We store the number of shifts made as well as the most significant 16 bits of the product to fully represent the product.

The process just described performs a 'normalised' floating point operation. The mantissa is stored to 16-bit accuracy, and makes full use of those 16 bits. The shift count is the exponent, and although in the above example it could only range from 5 to 31, it could also be stored as a 16-bit integer.

If all arithmetic on signal processor chips had to be carried out in the above form, they would be very poor real-time processors. The problem, as posed here, seldom arises in practice. It would usually be possible to save the product as a double-length integer if it is essential

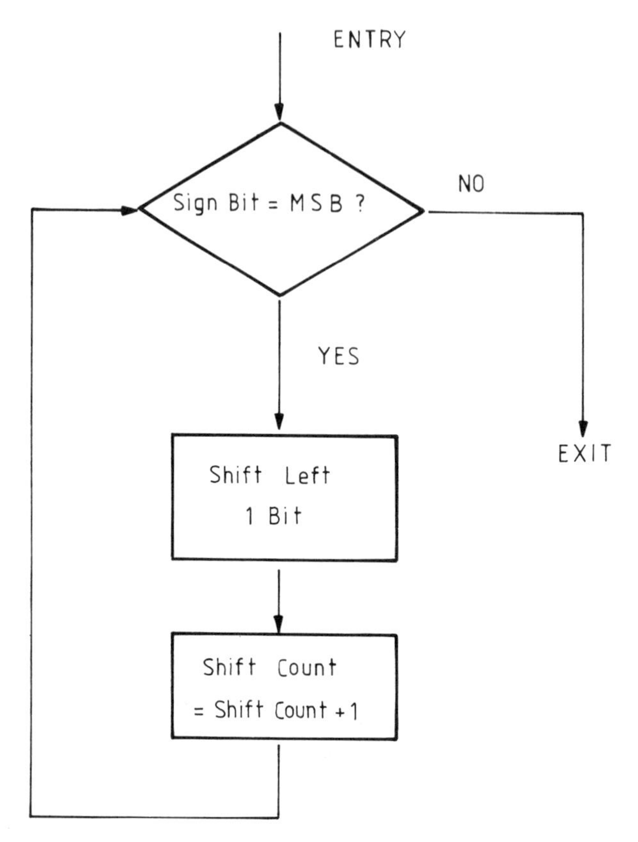

Flowchart 2.1 Floating point normalisation

to maintain the full accuracy. Nonetheless, the programmer must always be fully aware of the likely range of variables involved in a fixed point calculation, and he needs to keep careful track of where the (implied) 'point' lies. To put it another way, he must note, at every stage, the scaling used on his variables.

2.2 Four Rules Arithmetic

2.2.1

Little need be said about addition and subtraction, beyond noting that it suddenly becomes difficult for a floating point system. If two numbers to be added or subtracted have different exponents, these have to be made the same (a sort of de-normalising) before the operation.

2.2.2

Multiplication is the operation which has been greatly simplified in the signal processor chips. The earliest of the chips, the 2920, is the least convenient in this respect, but nonetheless, very high speed multiplication is possible. The techniques used depend on whether one of the numbers is fixed or not, and whether negative values are allowed. They are thoroughly covered in chapter 3. The hardware multipliers available in the 7720 and 320 deal with fixed point signal variables. In both cases the inputs are 16-bit 2's complement numbers, the product being a 31-bit 2's complement number, i.e. the inputs range from $+2^{15}-1$ to -2^{15}, and outputs range from $+2^{30}-1$ to -2^{30}. It is worth noting that multiplication of normalised floating point numbers is very easy. The mantissas are multiplied as fixed point numbers, the exponents are added, and re-normalisation will at most require one left shift and a corresponding increment to the exponent.

2.2.3

Division tends to be awkward, and is best avoided wherever possible. Division by a constant is easily replaced by multiplication using the reciprocal of the constant. Division by a variable requires a small program in all three of our devices. The algorithm usually used is exactly the long division process taught in schools for decimal arithmetic – with the proviso that the binary version is as usual much simpler. As a preliminary to division, both numbers are represented in sign and magnitude form, the division being carried out on the positive magnitudes, with complementing of the quotient if necessary after division. The quotient (QUO) is built up one bit at a time, starting at the most significant end. At each stage in the process the divisor (denominator DEN) is compared with the dividend (numerator NUM). If the divisor is bigger, a 0 is placed in the quotient, and the next comparison is between the dividend and the divisor shifted right one place. If the dividend is bigger, a 1 is placed in the quotient, and the next comparison involves a new dividend (which is the difference between the old dividend and the divisor) and the right-shifted divisor (see flowchart 2.2).

A variation on the theme includes left-shifting of the dividend instead of right-shifting of the divisor, and actual implementation depends very heavily on the particular device involved. For an example program using the 2920 see chapter 3 and program 3.2. Another fast device for implementing this type of division algorithm is

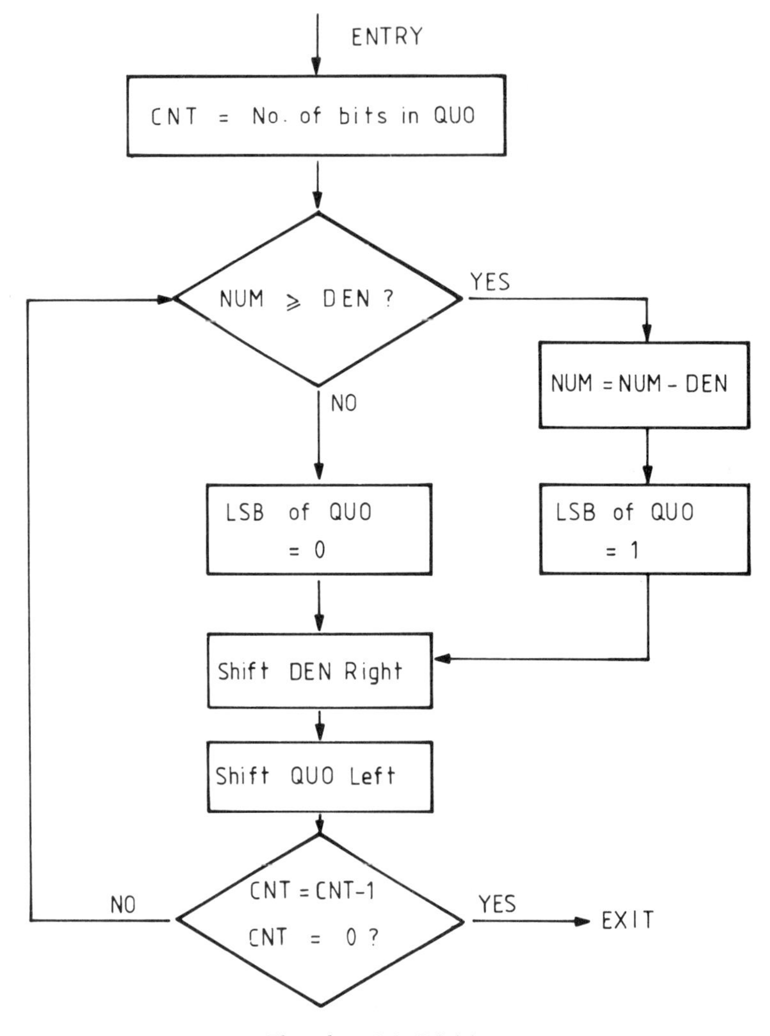

Flowchart 2.2 Division

the bit-slice ALU, the AMD 2903, in which one stage of the division can be accomplished in one instruction. A division routine for the TMS 320 is presented in ref. 2.1.

It has been found essential when writing division routines to try out examples of a particular problem on the particular device being used, and to do the operations manually, writing down the binary at each stage. It sounds laborious, but there are so many variations in scaling etc. that it is well worth while.

2.3 Transversal Filters

We begin by looking at the transversal filter because it is simple to implement, and the signal processor chips all suit its implementation very well. The block diagram of fig. 2.2 shows the filter structure. Following the usual notation, Z^{-1} represents a delay of one sample time, X_0 is the input sample, X_1 is the previous input etc., Y is the output, and A_0 to A_{N-1} are the filter coefficients.

$$Y = \sum_{k=0}^{N-1} A_k \cdot X_k$$

$N - 1$ is the 'order' of the filter. In general, the order refers to the number of delay elements used in the filter (flowchart 2.3). The transversal filter is by far the most commonly used example of the 'non-recursive' or 'finite impulse response' (FIR) filter type. It is not the only possible configuration for an FIR filter. A 'lattice' arrangement is another possible structure (fig. 2.3) which is used in some situations.

The impulse response of the transversal filter is trivial to observe. Starting from a situation where all X_k are zero, if a single sample of unity amplitude is put into the filter, and the input then returned to zero, the output will be the impulse response, or, more strictly speaking, the 'weighting sequence' of the filter. The sequence will be A_0, A_1, A_2 . . . A_{N-1}, and thereafter will be zero (hence the 'finite' impulse response).

This observation leads directly to a method of determining the

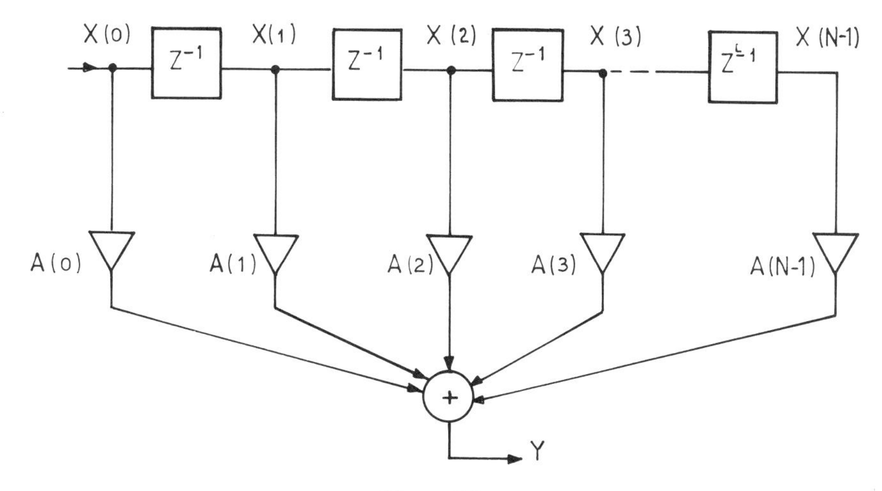

Figure 2.2

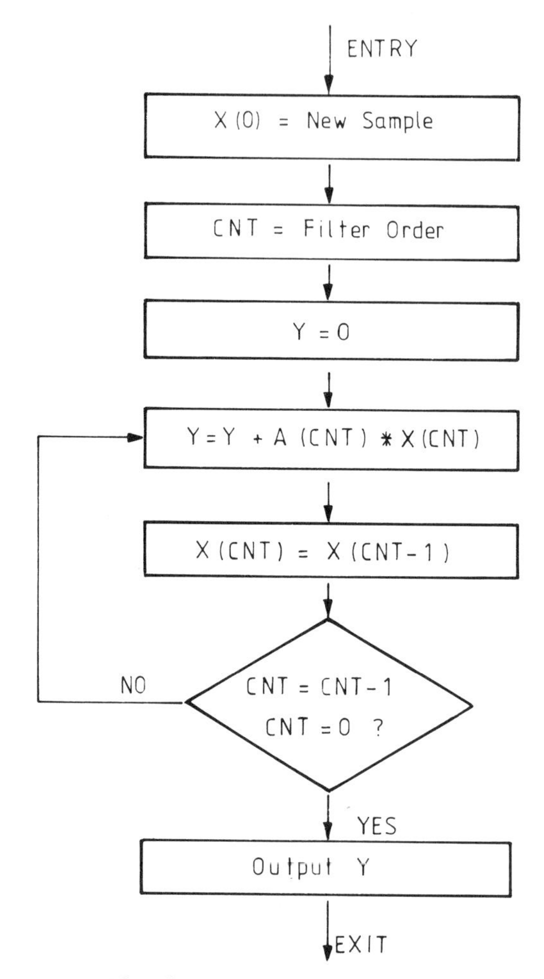

Flowchart 2.3 Transversal filter

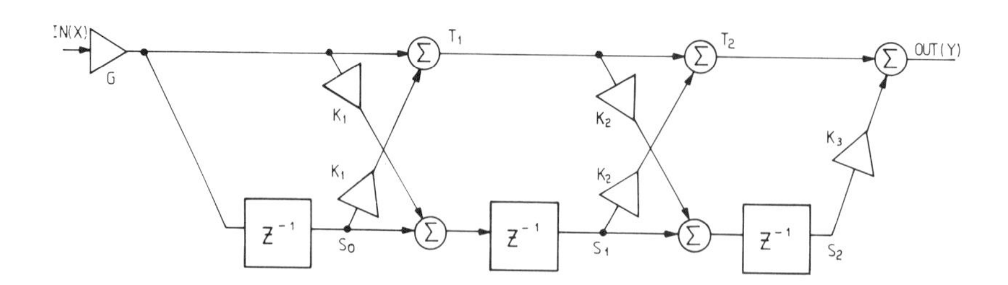

Figure 2.3 An FIR filter in lattice form

coefficient values which will be required in order to produce a fixed filter with any chosen frequency response. We use the fact that the sampled frequency response is the discrete Fourier transform (DFT) of the weighting sequence. So, the sampled frequency response is specified, both in magnitude and phase, and the inverse DFT is used to calculate the weighting sequence – hence the coefficients. This procedure is not normally done in real time, and standard routines to carry out the processing are available for most minicomputers. This theoretical approach results in general in complex values for the A_k. Purely real coefficients are normally required, and it can be shown that it is very straightforward to design a linear-phase filter with purely real coefficient. This is often a very desirable type of filter in that it produces minimum distortion of the waveshapes. The resulting coefficients are symmetrical about the central coefficient value ($A_{N/2}$) and tend to have their highest values around this position. This corresponds to an inherent delay of $N/2$ sample times.

The reader is recommended to standard texts on digital filters, such as Terrell (ref. 2.2, pp. 89–111) for a full treatment of this and other design methods, for fixed transversal filters.

In summary, the transversal filter is simple to implement, inherently stable, and straightforward to produce linear-phase designs. Its drawback is that longer filters are needed to achieve the same frequency selectivity as recursive designs.

2.4 Recursive Filters

Recursive filters are those involving feedback. When excited by an impulse they continue to produce a response, in theory, for ever, and so are called infinite impulse response (IIR) filters. Fig. 2.4 shows a very simple example of a recursive filter. It is a second-order filter (having two delay units), and its transfer function (H(z)) is

$$H(z) = G/(1 + B_1 z^{-1} + B_2 z^{-2})$$

This is easily seen if we look at the z transforms of the input IN(z), and output OUT(z), and multiply by z^{-1} for a delay of one sample period. The summation (to produce the output) can be written

$$OUT(z) - G \,.\, IN(z) - B_1 \,.\, OUT(z) \,.\, z^{-1} - B_2 \,.\, OUT(z) \,.\, z^{-2}$$

or $$OUT(z)\,(1 + B_1 z^{-1} + B_2 z^{-2}) = G \,.\, IN(z)$$

or $$OUT(z)/IN(z) = G/(1 + B_1 z^{-1} + B_2 z^{-2}) = H(z)$$

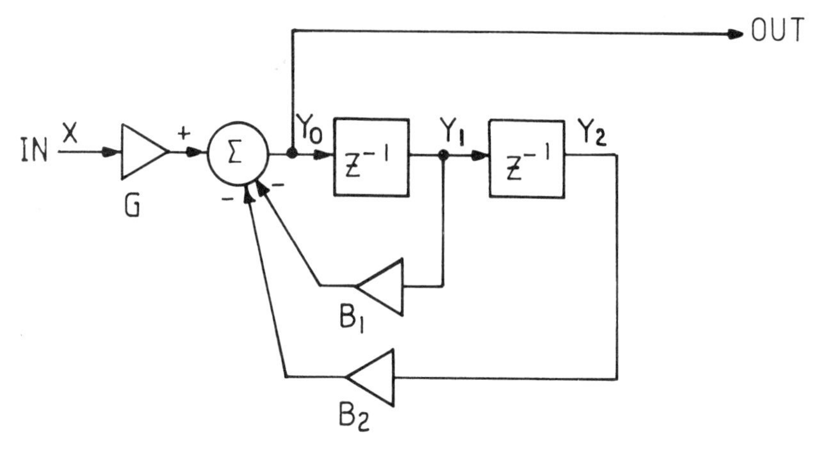

Figure 2.4 Second-order recursive filter

So, by using few 'rules of thumb' one can switch between the block diagram for a filter, and the transfer function in z notation.

When implementing a filter such as that in fig. 2.4 care is required in two particular areas. First of all, it is very easy to get the signs of the feedback coefficients muddled. In fig. 2.4 the feedback terms are shown as being subtracted. It may be more convenient, or a programmer may see it as more consistent, to add them. If this is the case, then of course the negative of the coefficients in our transfer function must be used. The second point concerns scaling of the coefficients. In a transversal filter, if all the coefficients are scaled by the same factor, the output is simply scaled by that factor. If, however, the values of B_1 and B_2 in fig. 2.4 are scaled by some factor, then this will completely change the frequency response of the filter. It is quite common in fixed point arithmetic to assume that the full range of integers represents the range -1.0 to $+1.0$. It can also happen that filter coefficients are required to be say, in the range -2.0 to $+2.0$. The scaling down by a factor of 2 needed to accommodate this wider range must be compensated for. A multiplication by 2 would be needed before the delayed output values are stored ready for the next sample to be filtered.

The transfer function above describes a system having a complex conjugate pair of poles. It represents a simple resonator. Higher order transfer functions can be used with larger numbers of poles (flowchart 2.4). For example, in the most commonly used model of the speech process, an all-pole transfer function is assumed. A 10th order filter is found to be about right for this model. High order filters implemented in the 'direct' form are very sensitive to calculation errors. The

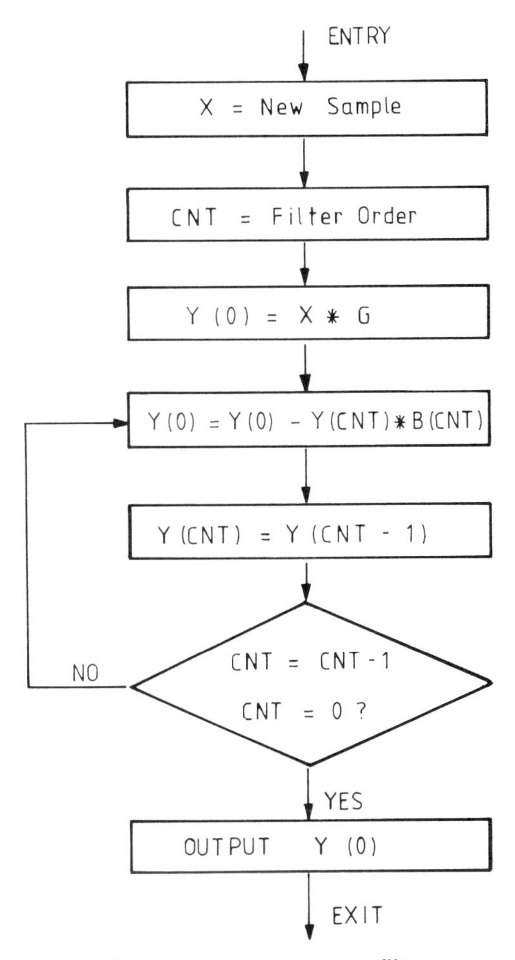

Flowchart 2.4 A recursive filter

summation of, in this case, ten products often has to produce only a small output value – the large positive contributions cancelling with large negative ones. Rather than attempt to work to very high accuracy it is better to split the filter into a cascade of shorter filters. In order to avoid any complex arithmetic, the obvious choice for the unit size is the second-order section, and so this is a very commonly used 'building block'. In the all-pole speech model, each block performs one resonance, and so the whole filter represents a cascade of resonances (or formants).

Another form of all pole filter is also often used as a speech production model. It is shown in fig. 2.5 and is a 'lattice' form. As compared with either a direct implementation or a cascade of second-order sections, it is complicated by the use of twice as many

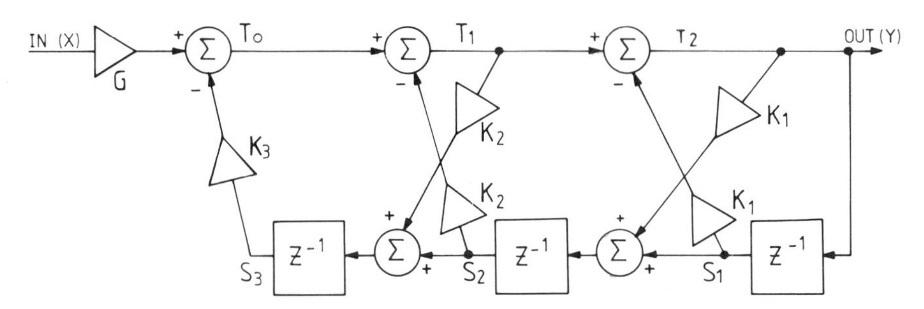

Figure 2.5 An IIR filter in lattice form

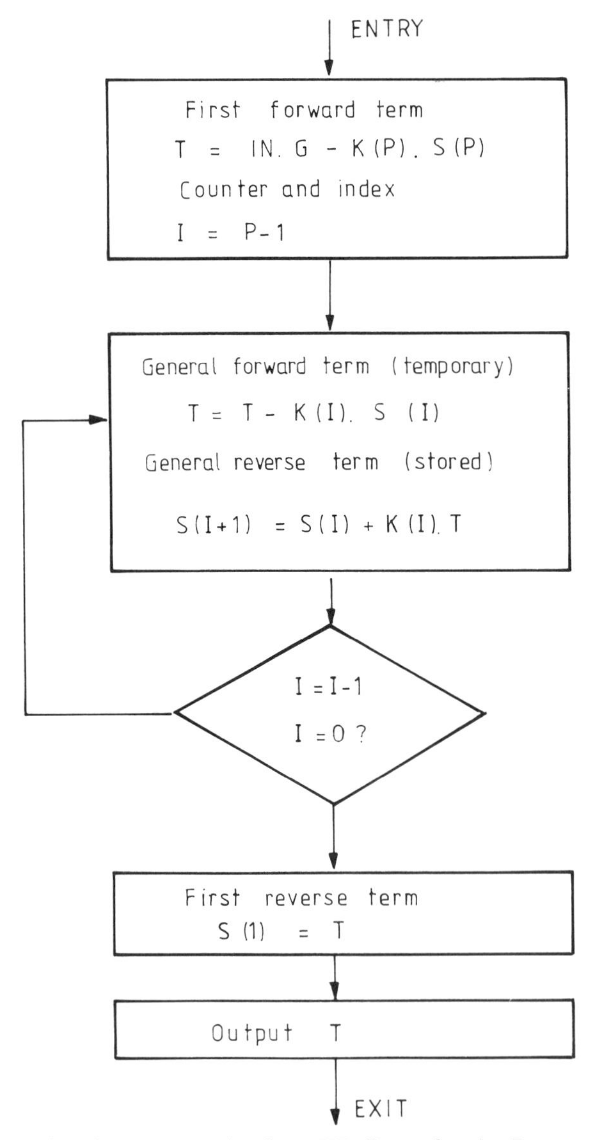

Flowchart 2.5 Lattice form IIR filter of order P

multiplications. The coefficients in this model are referred to as 'reflection coefficients', and have the advantage of being bounded within the range −1.0 to +1.0. They are also easy to calculate from a speech waveform. Summation is distributed through the model, and round-off errors in calculation are not as troublesome as in the direct form of the all-pole model (flowchart 2.5). This is the type of filter used by Texas Instruments in their LPC speech synthesiser chips TMS 5100, 5200 and 5220.

In general, filters having both poles and zeros are required, and the direct form of implementing such filters results in a block diagram such as fig. 2.6. Its transfer function is

$$H(z) = G \,.\, (A_0 + A_1 z^{-1} + A_2 z^{-2})/(1 + B_1 z^{-1} + B_2 z^{-2})$$

A_0 could be set to 1 by appropriate adjustments to G, A_1 and A_2. This is a 2 pole, 2 zero filter, and is the most commonly used building block in digital filters (flowchart 2.6). It is given the same 'biquad' or 'quartet' filter – or is simply called a second-order section. Routines to implement it are described for all three of our signal processor chips.

The derivation of the z-plane poles and zeros, and hence filter coefficients for recursive filters is covered in many texts, e.g. refs. 2.2 and 2.3. The methods used closely parallel those for analog filter design. Butterworth and Chebyshev approximations to the ideal lowpass filter can be implemented by following simple recipes. There are transformations by which low pass designs can be used to derive high pass, band-pass and band-stop filters.

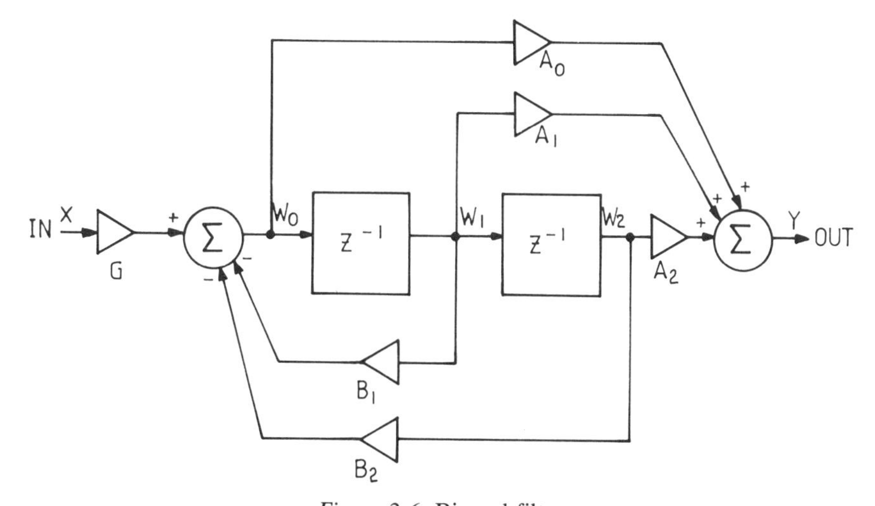

Figure 2.6 Biquad filter

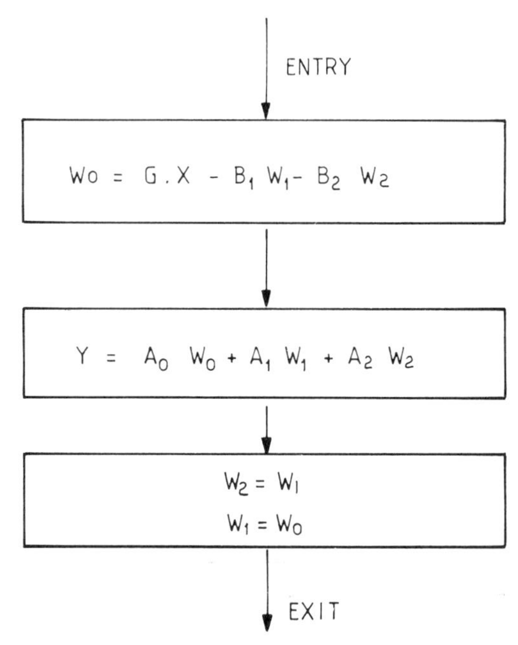

Flowchart 2.6 A biquad section

2.5 The Discrete Fourier Transform

Discovering the frequency content of a time waveform can be done in various ways, and it is important to select the right method for the application. If only certain frequencies are of interest, it will be best to set up filters to isolate these frequencies. The technique can be extended further, and a whole bank of filters can be set up to isolate a succession of frequencies. This technique is used in speech processing in the 'channel' vocoder. It is also used in the speech spectrograph, where a plot of frequency against time is produced, with picture intensity representing the amplitude at any given point. This method has the advantage that the filter bandwidths can be varied to suit the application. In the speech spectrograph, for instance, a narrow band analysis will reveal the harmonics of the fundamental (pitch) frequency, whereas a broader band analysis will show the resonances in the vocal tract (the formants).

There are circumstances when it is necessary to take a part of a waveform, and to transform it into the frequency domain. We could start by considering the Fourier transform. This is a way of converting a continuous waveform from the time domain to the frequency domain. It involves integration over infinite time – totally impracticable. We have to make approximations to be able to do anything. Firstly, we assume that the waveform is zero outside some range, and secondly, that samples of the waveform can be taken to adequately represent it (the usual Nyquist rate applies here). We can then replace integration by a set of additions.

Another way of looking at the problem is to take our chunk of waveform, and to consider it as a single cycle of a periodic waveform. This could be just the thing to do if our chunk was a single pitch period isolated from a speech waveform, for instance. This waveform can be represented by a Fourier series, the representation being contained in the coefficients of sine and cosine terms. The calculation of the coefficients again involves integration, this time over a finite time interval, and once again, we could approximate to this integral by a summation.

It turns out that both of these approaches lead us to the same approximation, a transformation called the discrete Fourier transform (DFT). This is a transformation which can be done on a computer, involving multiplication and summation. It takes in a finite length sequence of numbers, which could in general be complex numbers (although for samples of a time series they would be real) and produces a second finite series of complex numbers (if these are frequency values, they could be converted to magnitude and phase information). If we consider N input numbers

$$x_0, x_1, x_2 \dots x_k \dots x_{N-1}$$

and the N output numbers are

$$X_0, X_1, X_2 \dots X_l \dots X_{N-1}$$

then the transform to give any one of the outputs is:

$$X_l = \sum_{k=0}^{N-1} x_k \cdot W_{lk}$$

where W_{lk} is one of an array of complex constants. These constants are defined by the formula

$$W_{lk} = \exp(-j \cdot 2\pi lk/N)$$

Their real and imaginary parts are therefore

$$\text{Re}\,(W_{lk}) = \cos(-2\pi lk/N)$$

and $$\text{Im}\ (W_{lk}) = \sin(-2\pi lk/N)$$

When we consider real-time processing it is unlikely that we would attempt to calculate these sines and cosines. They would be pre-stored in a table.

The computing resources needed therefore include a fairly large store for constants, and for each output value we need to perform N multiplications between a complex constant and a variable which may be either complex or, for samples of a time waveform, real.

The use of a signal processor chip for these multiplications makes this direct approach to frequency analysis a possibility in its own right. N can be chosen quite freely – an advantage not shared by faster methods of computation. We consider these faster methods in the next section.

2.6 The Fast Fourier Transform (FFT)

The FFT algorithms are simply fast methods of performing the DFT. They were first described by Cooley and Tukey (ref. 2.4) in 1965. There are two methods which lead to identical results. One is called 'decimation in time' and the other is called 'decimation in frequency'. All that this means is that the order of the input samples (for decimation in time) is rearranged to make the program flow easier; or alternatively, the rearranging can be done on the output samples after calculation (decimation in frequency). The words 'shuffling', 'twiddling', 'unscrambling', etc. are variously used to describe the rearrangement of the order, which is in fact a very simple and systematic rearrangement.

The method relies on splitting up the DFT operation on the N values $x_0, x_1 \ldots x_{N-1}$, into two smaller DFT operations, each done on $N/2$ of the values. The results of these smaller DFTs are combined, using an appropriate formula, to give the larger DFT result. If $N/2$ is itself even, the operation can be repeated again in order to calculate the smaller DFTs. In fact, if N is a power of 2, the splitting up can go on until eventually the DFT has to be calculated on a sequence containing just one value. The DFT of one number is the number itself, and so that is trivial. The only problem in the whole procedure is therefore the relating of the two $N/2$ point DFTs to the whole N point DFT. The derivation does not involve much mathematics, but here we will confine ourselves to stating the result and referring the reader to any standard text should he wish to follow the maths. Our result describes the 'decimation in time' approach. The input values

$$x_0, x_1, x_2 \ldots x_{N-1}$$

are divided into two groups

$$x_0, x_2, x_4 \ldots x_{N-2}$$

and

$$x_1, x_3, x_5 \ldots x_{N-1}$$

these are then re-labelled

$$y_0, y_1, y_2 \ldots y_{(N/2)-1}$$

and

$$z_0, z_1, z_2 \ldots z_{(N/2)-1}$$

The DFTs of the new sequence are called

$$Y_0, Y_1, Y_2 \ldots Y_l \ldots Y_{(N/2)-1}$$

and

$$Z_0, Z_1, Z_2 \ldots Z_l \ldots Z_{(N/2)-1}$$

then

$$X_l = Y_l + W_l \cdot Z_l$$

and

$$X_{l+N/2} = Y_l - W_l \cdot Z_l$$

where the W_l values are complex constants given by the formula

$$W_l = \exp(-j \cdot 2\pi l/N)$$

The real and imaginary parts of W_l are

$$Re(W_l) = \cos(-2\pi l/N)$$

$$Im(W_l) = \sin(-2\pi l/N)$$

The case of $N = 8$ is a good one to illustrate the procedure, and the above equations can be represented diagrammatically, as in fig. 2.7. The blocks marked 4-point DFT can each be expanded using the same approach, shown in fig. 2.8. Finally, blocks marked 2-point DFT can each be expanded, as shown in fig. 2.9.

The whole combined operation, starting with the 2-point DFTs,

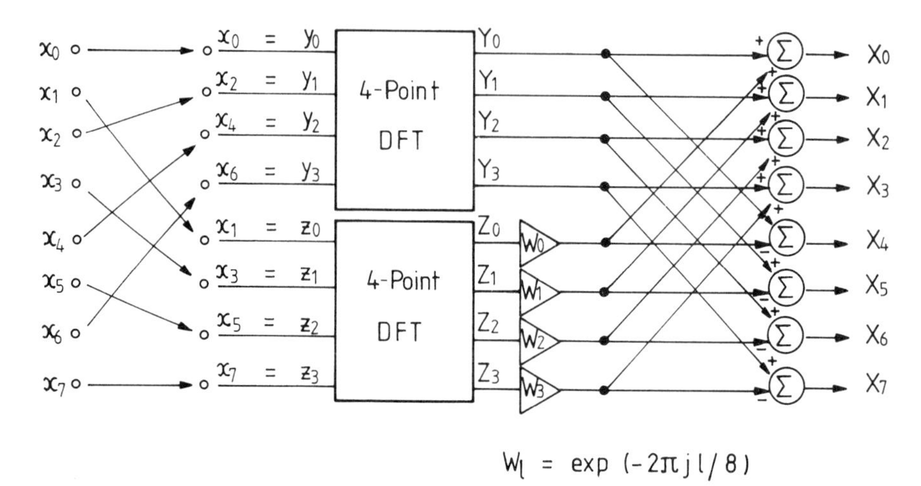

Figure 2.7 8-point DFT split into 4-point DFTs

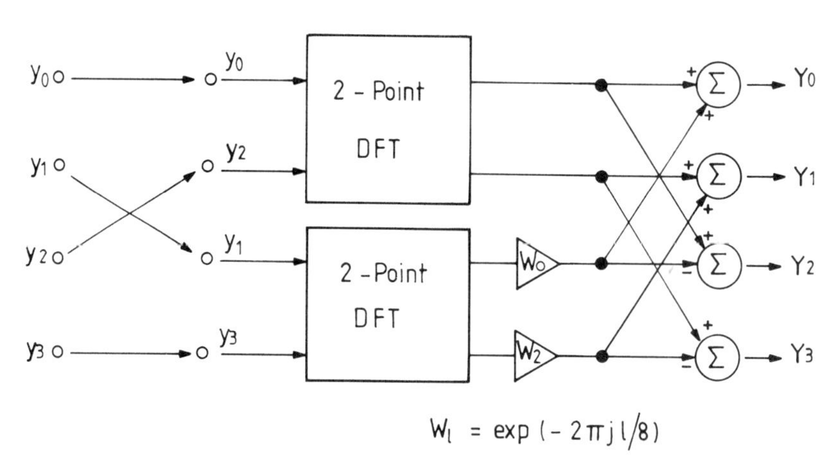

Figure 2.8 4-point DFT split into 2-point DFTs

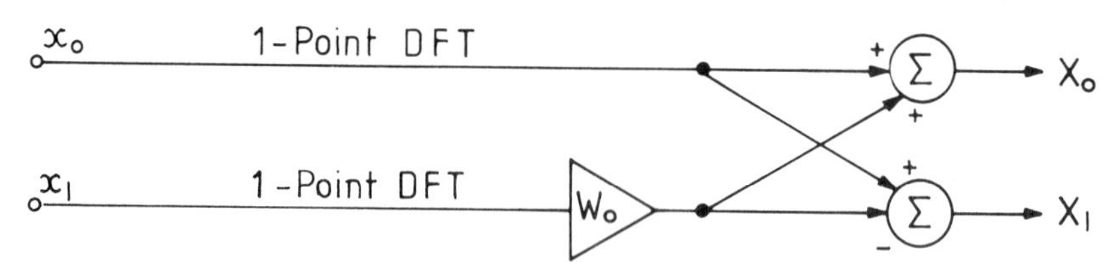

Figure 2.9 2-point DFT split into 1-point DFTs

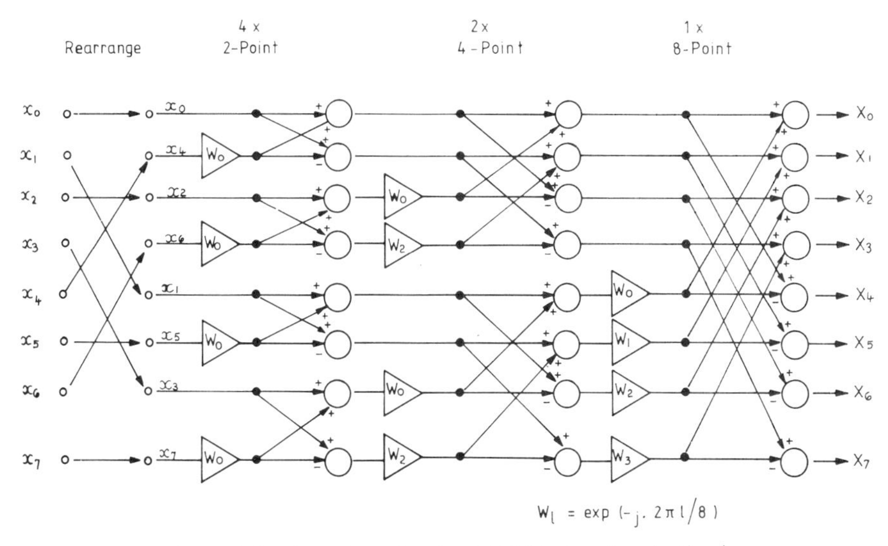

Figure 2.10 Complete 8-point FFT (decimation in time)

using them to provide input to the 4-point operations, which in turn give the inputs to the 8-point operation, is shown in fig. 2.10. This is a typical decimation-in-time FFT flow diagram. The points to note are firstly that the operation which is performed over and over again is that shown in fig. 2.11, known as the FFT 'Butterfly'. Secondly, the rearranging of the data before the whole operation can best be described by looking at the binary values of the x subscripts (addresses in computer terms). The order of the binary digits is simply reversed (fig. 2.12).

$$W_l = \exp(-j \cdot 2\pi l/N)$$

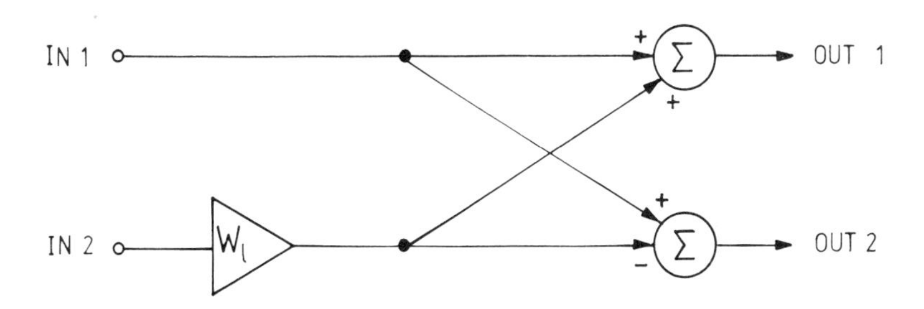

Figure 2.11 FFT butterfly (decimation in time)

Decimal subscript (Variable address)	Binary	
0	000	
4	100	
2	010	Order of binary
6	110	digits, if reversed,
1	001	gives straight binary
5	101	sequence
3	011	
7	111	

Order reversal applies generally to 2^N block size

Figure 2.12 Input data ordering for FFT by decimation in time

The calculation saving over the direct DFT calculation can be considerable when larger block sizes are considered. Many more variations exist. The decimation in frequency approach arranges the calculations differently.

Chapters 4 and 5 give example programs for the NEC 7720 and TMS 320 respectively, using the decimation in frequency approach. The butterfly operation is different to that of the decimation in time method. It is illustrated in fig. 2.13. There is, of course, great similarity to the decimation in time method, and the calculation complexity of the two approaches is the same. Fig. 2.14 illustrates the calculation ordering and data re-arranging for an 8-point FFT by decimation in frequency.

Our description is the radix-2 approach – the smallest calculation block being the butterfly with 2 inputs and 2 outputs. Radix-4 and radix-8 methods can be even more efficient. As with the DFT, the FFT can take advantage of the case where the input values are all real.

$$W_R = \exp(j.2\pi R/N) = \cos(2\pi R/N) - j\sin(2\pi R/N)$$

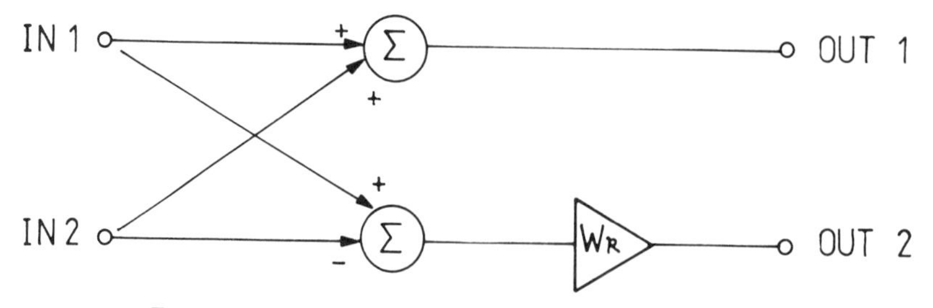

Figure 2.13 The FFT butterfly (decimation in frequency)

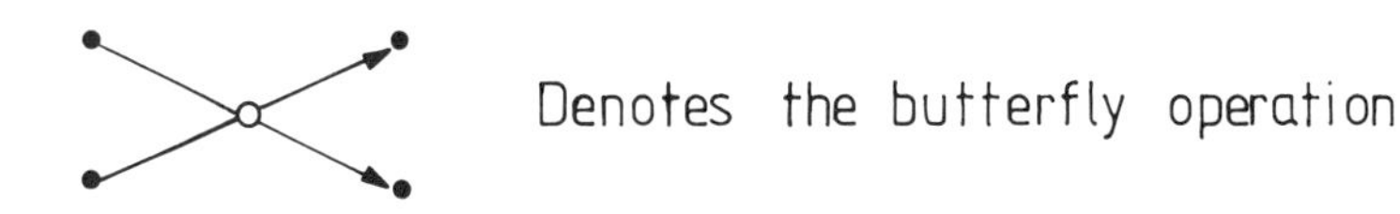

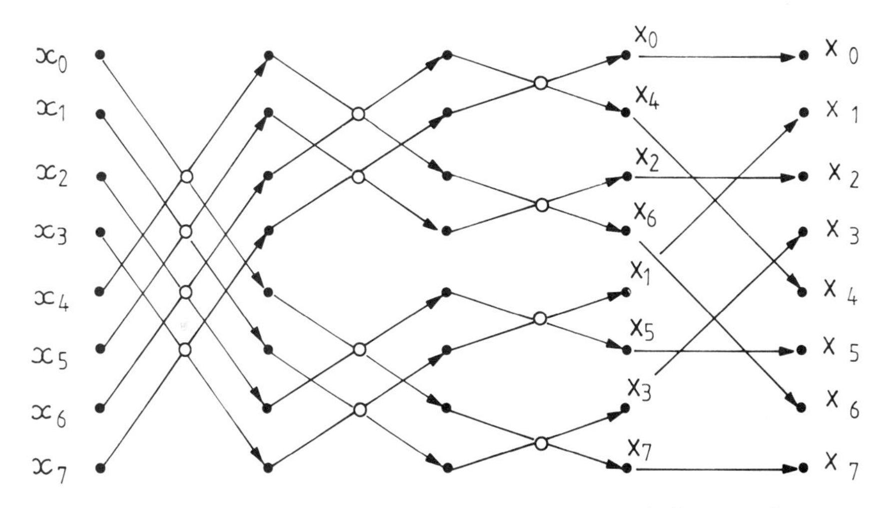

Figure 2.14 Complete 8-point FFT (decimation in frequency)

2.7 Linear Prediction

Linear predictive coding (LPC) is an indirect form of spectral analysis. It is a method of estimating the coefficients of a digital filter, the transfer function of the filter containing the spectral information. The most frequent method used derives the fundamental spectral information from estimates of autocorrelation coefficients, and in fact the calculation of these coefficients is the most time consuming part of the procedure. We will consider the estimation of autocorrelation coefficients first, and then the methods by which filter coefficients are derived from them.

An autocorrelation coefficient (R_m) is defined as the mean value of the product of two signal values.

$$R_m = \text{mean of } X_n * X_{n-m}$$

and so $\quad R_0 = \text{mean square signal value.}$

When making estimates of these means on signals such as the speech signal (the most commonly used signal in current LPC work) we have the problem that we are dealing with a signal of changing spectral content. The number of samples over which the means are calculated is therefore a careful balance between a large enough number to provide a reliable estimate, and a small enough number to ensure adequate time resolution. For speech, a time interval of 20 to 30 ms is generally found to be best. If sampling is at 10 kHz, this might correspond to a block of, say, 256 data points. The time resolution can be improved by the use of overlapping windows, and a shaping function can be applied to the data windows in the same way as for any other form of spectral analysis.

A problem which can arise when using small systems based on signal processor chips is a shortage of RAM. One method of overcoming such a shortage is to form a running estimate of each of the autocorrelation coefficients. Assume that ten coefficients are to be estimated (a typical number for speech work). A shift register arrangement of ten sample values can store all the samples needed for the ten products, each product being a one sample estimate of the coefficient. The one sample estimate can then be low-pass filtered to provide a continuous, smooth, estimate of each coefficient. There is, of course, a penalty to pay in the calculation time associated with the smoothing filter, but this can be a very simple one – just a first- or second-order recursive structure. This approach has a great deal to recommend it. Low storage, great flexibility in the smoothing/time resolution tradeoff, and an estimate available at any chosen time (after each sample if need be). A flowchart is presented as flowchart 2.7.

It was noted earlier, but is worth repeating, that the product of two signal values requires twice the number of bits in order to accommodate its increased dynamic range. This often involves working to 32-bit accuracy when performing autocorrelation estimates on 12 or 14-bit signal samples.

No attempt will be made here to explain the mathematical basis for the transformation from autocorrelation coefficients to filter coefficients. Interested readers should start with ref. 2.5. The principle underlying the method is to deduce the coefficients needed for a transversal filter (all zero filter) which will 'whiten' the signal (produce a flat power spectrum). This leads to a set of equations (or a matrix equation) which involves the filter coefficients and the autocorrelation coefficients. There is a very fast and straightforward recursive algorithm by which this set of equations can be solved, which in its original form was due to Levinson (ref. 2.6). The algorithm produces not only the coefficients for the transversal filter, but simultaneously derives the reflection coefficients for a corresponding lattice filter.

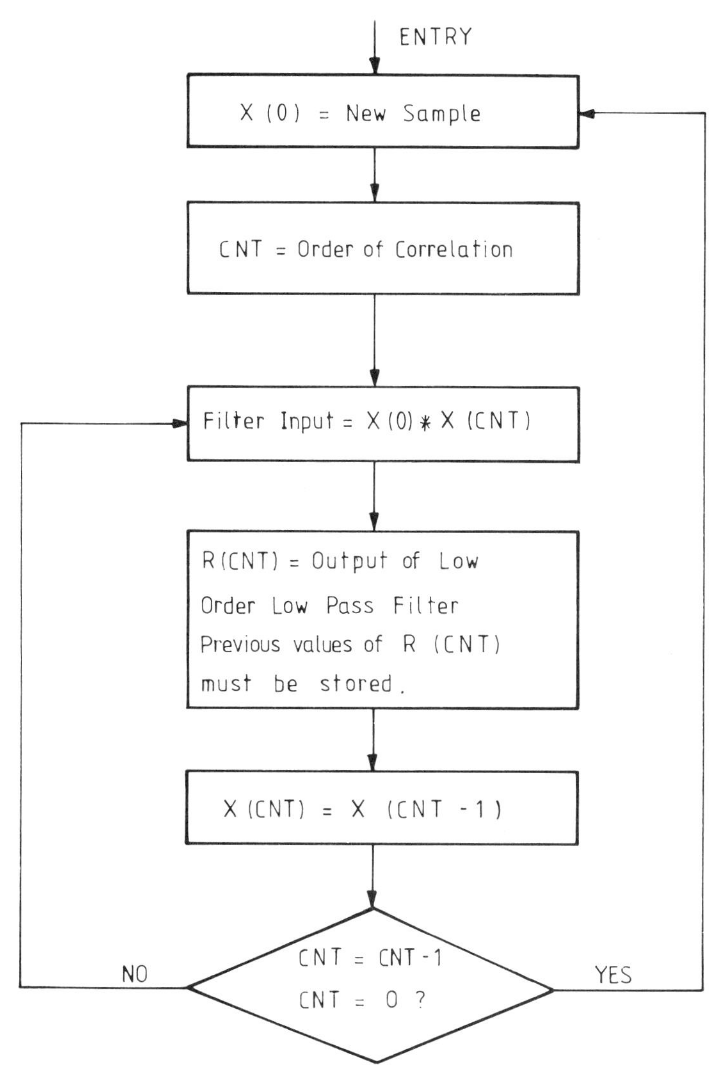

Flowchart 2.7 Running estimation of autocorrelation coefficients – R(CNT) represents an autocorrelation coefficient

Many authors have refined the recursion, and the most useful version of it for fixed point arithmetic, credited to LeRoux and Gueguen (ref. 2.7) is presented here. This version gives only the reflection coefficients, but has the great advantage that these, and all intermediate variables, are bounded to the range -1.0 to $+1.0$. Flowchart 2.8 shows the LeRoux–Gueguen algorithm. Where an 'index' is used, the operations actually performed are alterations of address pointers. Two temporary arrays, EP and EN are used, each requiring P variables, where P is the order of the analysis. Two loops,

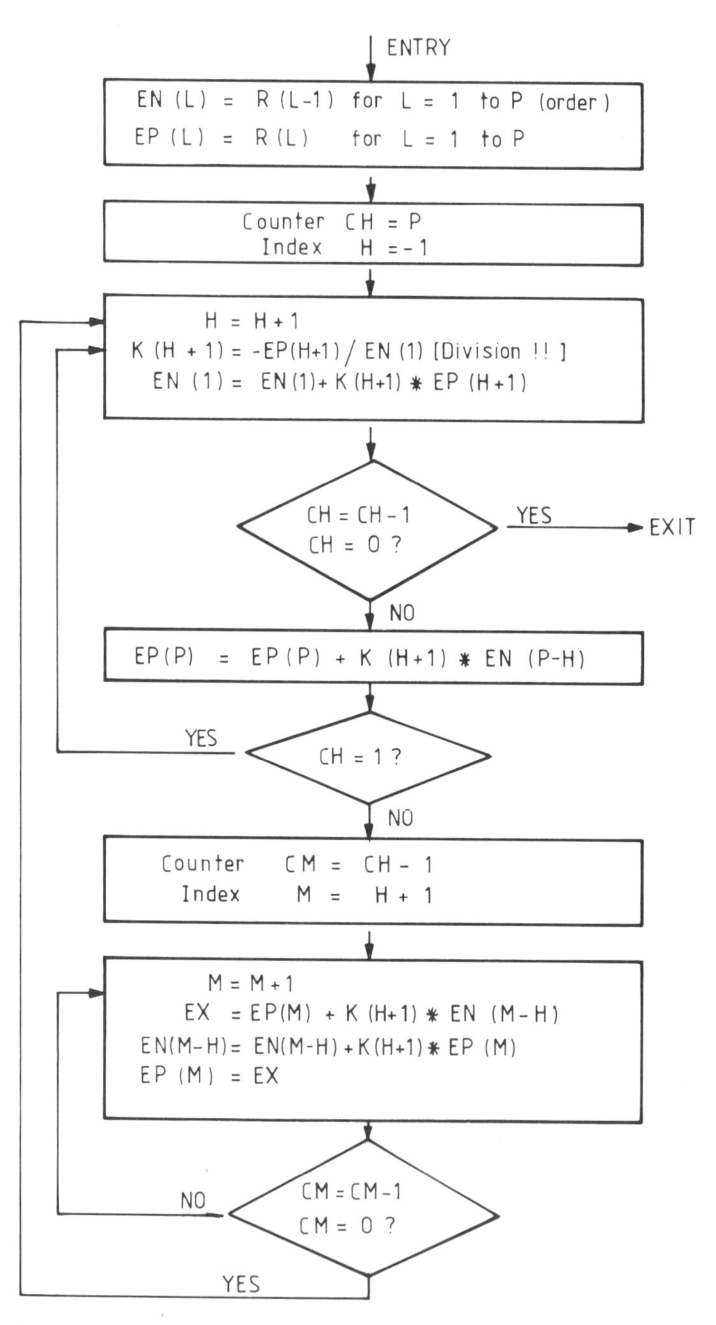

Flowchart 2.8 Autocorrelation coefficients converted to reflection coefficients

one nested inside the other, are used for the calculation of reflection coefficients. Although these are in theory bounded to the range −1 to +1, errors in the estimates of autocorrelation coefficients can lead to erroneous values beyond this range, and it is sensible to check for overflow from the range. When this occurs, it is advisable to disregard the results and revert to the previous set of reflection coefficients. The only difficult feature of the calculation is the unavoidable division

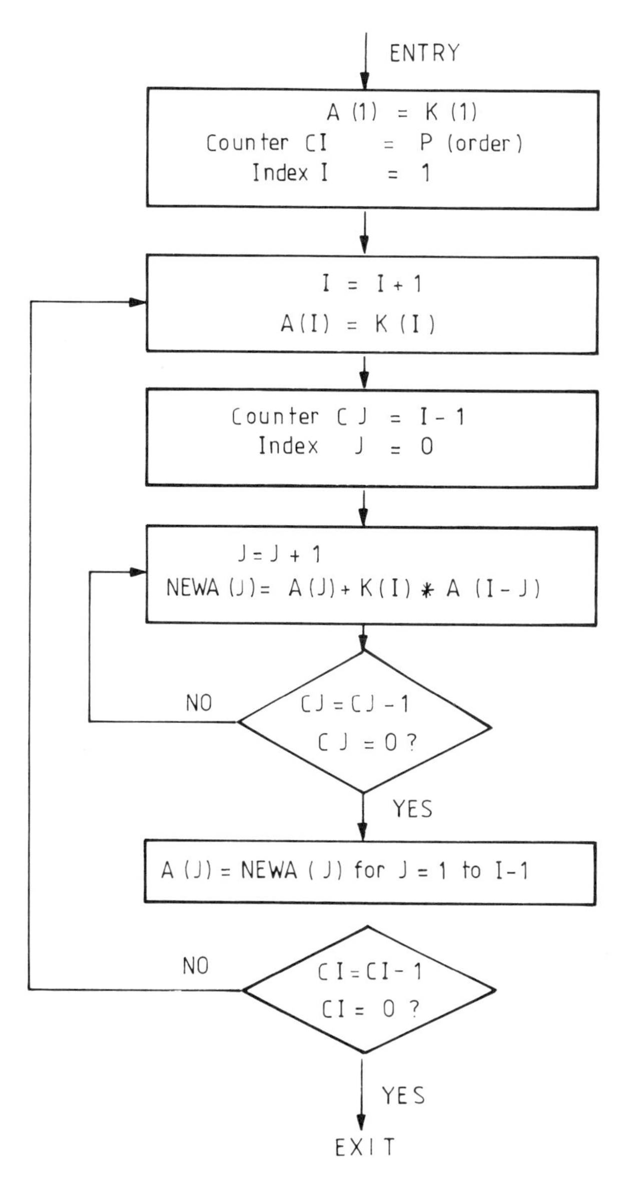

Flowchart 2.9 Reflection coefficients converted to transversal filter coefficients

needed for each reflection coefficient. In real-time speech processing this algorithm would proceed as a 'background' calculation. The operations required to deal with each sample would periodically interrupt it, requiring some careful attention to details of timing. A subsequent recursion can be used if the coefficients of the transversal filter are required, thus isolating the occurrence of numbers outside the -1.0 to $+1.0$ range (flowchart 2.9).

In an LPC speech vocoder, the filter coefficients are transmitted, and speech synthesis at the receiver is carried out on an inverse of the whitening filter. The inverse of the all-zero transversal filter is an all-pole recursive structure, which in its direct form will use the same coefficients as the transversal filter. As mentioned in section 2.4, it is better to use the lattice structure, and for this the reflection coefficients are available.

In spite of its central role, it is rarely necessary to actually implement the transversal filter, even in a real-time vocoder, since the process of determining the coefficients is all that is required. Only in the cases where further coding on the result of the filtering (called the 'residual' signal) is required, is it necessary to implement it. An example of such a form of vocoder is the 'residual excited linear predictive' (RELP) vocoder (ref. 2.8).

2.8 Power Series Expansions

The majority of standard functions in a real-time processing system would be derived from look-up tables. There is, however, a reasonably fast alternative in the form of a converging power series expansion which can be used in many circumstances, and might be essential if ROM is limited. The general form of the calculation would be

$$f(x) = a_0 + a_1x + a_2x^2 \ldots a_nx^n$$

Flowchart 2.10 shows a suitable algorithm to perform this operation. Each term requires two multiplication operations, one to create the power of x, starting with the previous power of x in the series, and a second to include the coefficient. x will usually be in the range -1.0 to $+1.0$, and clearly, the more rapidly the series converges, the more appropriate it is to use this technique. In the flowchart, the function value is denoted by FUNC, the power of x by POX, and coefficients by A(I).

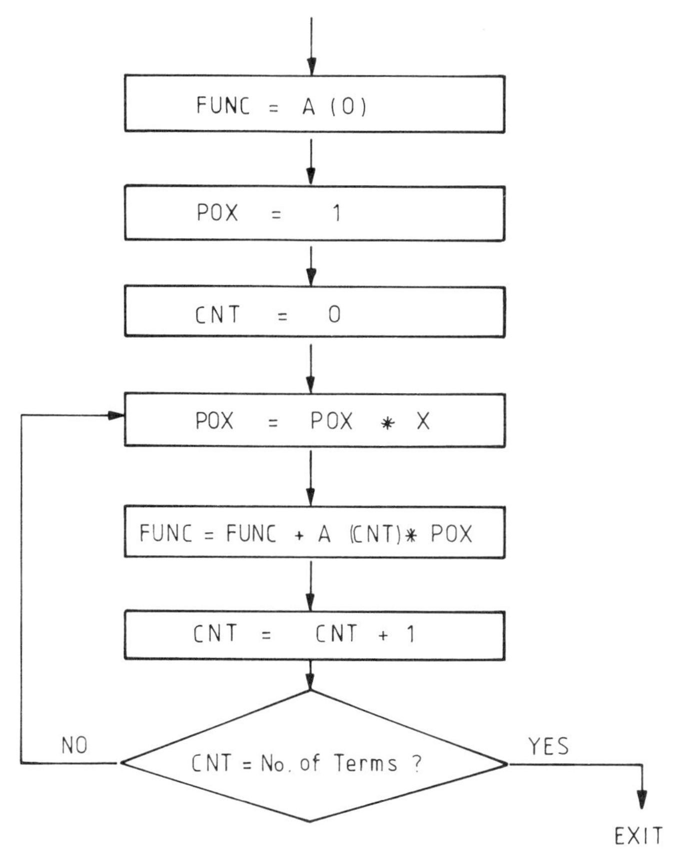

Flowchart 2.10 A power series expansion

2.9 Pattern Matching

Pattern matching is very intensive in its use of computer arithmetic. This has been one of the many factors which has limited the widespread use of pattern recognition. There are many other, more fundamental, limitations in most pattern recognition areas, but the signal processor chips have a role to play in reducing some of the computational difficulties. One of the first mask-programmed versions of the NEC 7720, the NEC 7762, is arranged to perform pattern comparison for speech recognition.

In the past, it has often been expedient to express a pattern as a binary vector of descriptors, or features. This was done so that a simple comparison could be made between a stored binary pattern and the unknown using the exclusive-OR operation. A much more common requirement is to compare feature vectors of analog measurements. If an unknown pattern vector has two components (U_1, U_2) and is to be compared with a stored pattern (K_1, K_2) the most commonly used method of comparison is to calculate the squared Euclidean distance (DIST) between the two.

$$\text{By Pythagoras: DIST} = (U_1 - K_1)^2 + (U_2 - K_2)^2$$

Similar calculations are performed between the unknown and other stored patterns, and the one to which the unknown is nearest can be determined. Since the problem is merely to find the nearest pattern, the squared distance measurement serves just as well as the distance – avoiding the square root calculation.

In general terms the feature vectors will have more than two components. The Euclidean distance metric generalises very easily to *n* dimensions, the only problem lying with the human imagination as it contemplates *n*-dimensional hyperspace.

In general $$\text{DIST} = \sum_{i=1}^{n} (U_i - K_i)^2$$

Once again, this calculation intensively uses the multiply/accumulate

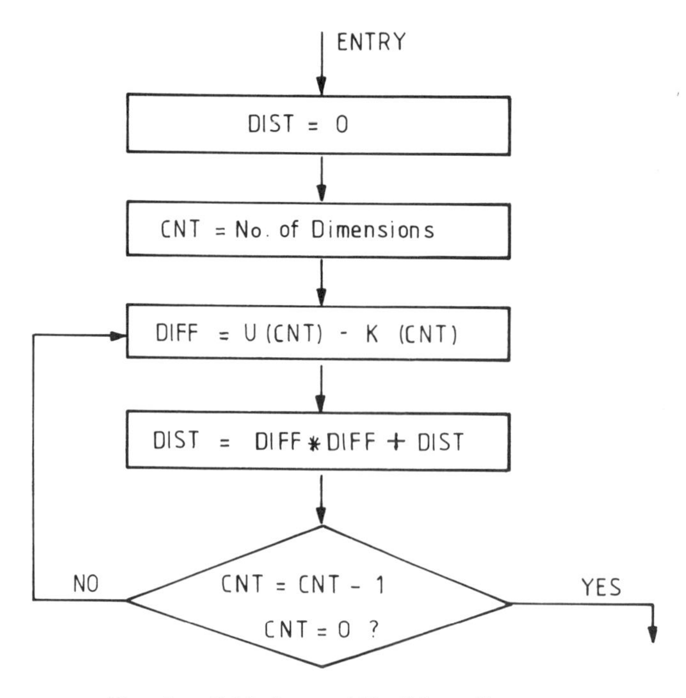

Flowchart 2.11 Squared Euclidean distance

operation, and so is well matched to implementation on a signal processor chip (flowchart 2.11).

References

2.1 Morris, L.R. *Digital Signal Processing Software* DSPS inc.
2.2 Terrell, T.J. (1980) *Introduction to Digital Filters* Macmillan.
2.3 Ackroyd, M.H. (1973) *Digital Filters* Butterworths.
2.4 Cooley, J.W. and Tukey, J.W. (1965) 'An Algorithm for the Machine Calculation of Complex Fourier Series' *Mathematics of Computing* **19** 297–301.
2.5 Makhoul, J. (1975) 'Linear Prediction – A Tutorial Review' *Proc. IEEE* **63** pp. 561–580.
2.6 Levinson, N. (1947) 'The Wiener RMS error-criterion in filter design and prediction' *J. Math Phys.* **25** pp 261–278.
2.7 LeRoux, J. and Gueguen, G. (1977) 'A Fixed Point Computation of Partial Correlation Coefficients' *IEEE International Conference on Acoustics, Speech and Signal Processing.*
2.8 Un and Magill (1975) 'The Residual-excited Linear Prediction Vocoder with Transmission Rate below 9.6 kbits/s *IEEE Trans* COM-23, No. 12, December 1975.

CHAPTER 3
THE INTEL 2920

J. Rittenhouse, Intel Corporation

3.1 Introduction

The Intel 2920 is a single chip, programmable signal processing device intended as a direct replacement for analog subsystems. The 2920 has on-chip program and scratch pad memory, D/A and A/D converters, sample-and-hold circuitry, and a high speed digital processor. As such, the 2920 acts as a complete, stand-alone sampled data system. Only anti-aliasing and rcconstruction filters, if needed by the application, and a few external components are required.

The 2920 can be used in a wide range of real-time signal processing applications, including modems and other communications applications, sonar and sound processing, speech processing, guidance and control, complex filtering, and many other applications where the signal bandwidth is in the audio range. Because the 2920 has four inputs and eight outputs, it can be used to implement a fairly complicated subsystem with multiple I/O requirements, or several less complex, independent functions.

The 2920 has EPROM program memory, and is therefore user programmable and erasable. The 2921 is the mask programmable ROM version of the signal processor. The following material will refer to the 2920 with the understanding that the comments also apply to the 2921. There are some differences in output drive capability, input impedances, and input/output sequences between the two devices.

Examples are given using the 2920 for illustrative purposes. In either case, the current data sheet should be consulted for up-to-date information on this type of data.

3.2 2920 Description

3.2.1 Overview

A block diagram of the 2920 is shown in fig. 3.1. This block diagram can be divided into three major sections: the program storage area, the digital processor, and the analog I/O section. These sections will be described briefly first and then covered in some detail.

The program memory section of the 2920 consists of an instruction clock generator, program sequence counter, and 192 locations of user memory (ROM for the 2921 and EPROM for the 2920). Each location holds a 24-bit instruction which is divided into several fields. These fields control the various sections of the digital processing and analog portions of the 2920.

The digital processing section of the 2920 consists of a 40 word by 25-bit RAM with two ports, a scaler (shifter), and an arithmetic and logic unit. The DAR (digital/analog register) performs the interface between the digital and analog sections.

The analog section of the 2920 interfaces the digital processor to the outside world, and performs the A/D, D/A and sample-and-hold functions. This section consists of a four input multiplexer, input sample-and-hold, D/A converter, a comparator, and an output demultiplexer with eight sample-and-hold circuits.

The basic operation of the 2920 can best be understood by following the flow of a signal through the 2920 as it is being processed. Under control of the program memory, one of the input channels is selected, and a sample of the selected signal stored on the sample-and-hold capacitor. The signal sample is then converted to a digital word, using

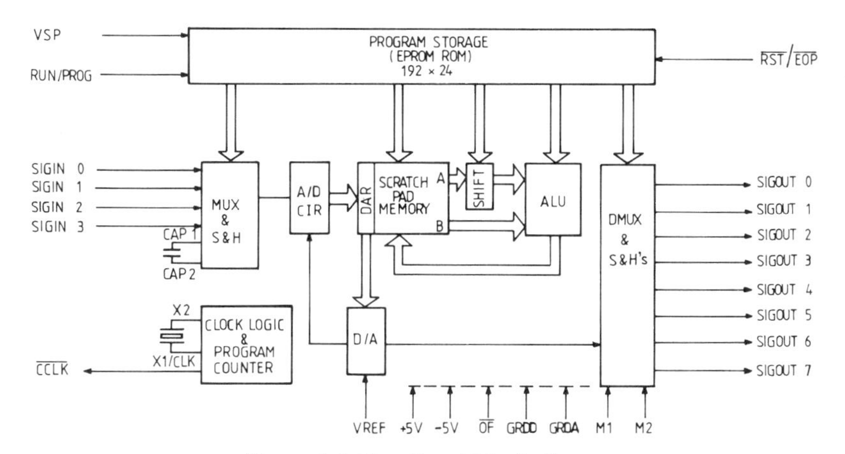

Figure 3.1 Functional block diagram

a successive approximation technique, with up to 9-bit accuracy (8 amplitude bits plus sign). The result is stored in the DAR. The value in the DAR can then be moved to the scratch pad memory for further processing, or operated on directly.

The digital section of the 2920 will be operating simultaneously with the analog section (using previously acquired samples of course). For example, a multiple stage filter could be realised using the digital processor during a 9-bit A/D conversion. The ALU operates on two 25-bit values accessed simultaneously from the RAM. The value read from the A port is passed through a binary shifter. The scaled A value and the unscaled B value are then processed by the ALU using the arithmetic or logic instruction specified by the program memory. The 25-bit result is written back to the B address location of the RAM.

To output the processed signal, the nine most significant bits of a computed sample value are loaded into the DAR. The DAR drives the D/A converter, the output of which can be routed under software control to any one of the eight output sample-and-hold circuits.

What makes the 2920 fast enough for real-time processing of analog signals is that an analog operation, dual memory fetch, binary shift, ALU execution, and write back to RAM can all take place in one instruction cycle. With a 10 MHz clock, an instruction cycle is 400 ns.

3.2.2. The ROM Section

The ROM section contains 4608-bits arranged as 192 words of 24-bits each. Each 24-bit instruction consists of five fields, one of which controls the analog section, while the remaining four control the digital section. There are two 6-bit fields which address the A and B locations in the scratch pad RAM, a 4-bit field to control the shifter, and a 3-bit field which identifies the op-code for the ALU.

The instructions in ROM are executed sequentially. There is no program branching. Control returns to instruction 0 after instruction 191 is executed, or when an EOP instruction is encountered. All instructions execute in the same amount of time, with each instruction requiring four master clock cycles. This results in a constant sample rate for the 2920 which depends only on clock rate and program length.

The ROM fetch/execute cycle is pipelined four deep. This is invisible to the user, except in the case of the EOP instruction, which must be located at an address which is divisible by four, e.g., location 0, 4, . . . , 188. The three instructions following the EOP will be executed. An open drain output, $\overline{\text{CCLK}}$, indicates the beginning of an instruction fetch cycle with a low transition. This signal remains low during execution of the first instruction of the cycle. The $\overline{\text{CCLK}}$ output completes one cycle for every sixteen master clock cycles.

An active low, open drain signal, $\overline{\text{RST/EOP}}$, is available as an indication that an EOP instruction has been executed. This output can drive one LSTTL load. A pull-up resistor is required for proper operation of the EOP instruction. The $\overline{\text{RST}}$ pin also functions as an input and can be pulled low to force the 2920 to reset its program counter. This is useful when it is necessary to synchronise a 2920 with some external event. An OR-tied connection may be used with this signal. Internally the EOP and external reset are equivalent functions, so the external reset must conform to the same placement rules as for the EOP instruction. The $\overline{\text{CCLK}}$ output can be used to strobe the reset signal. The 2920 will continuously loop through the first four instructions if RST is held low.

3.2.3 Digital Processing Unit

The digital processing subsystem of the 2920 has a typical signal processing architecture, allowing dual operand fetch, ALU execution, and write back to memory in the same instruction cycle. The digital processor uses 25-bit two's complement arithmetic. All numbers are treated as fractions, with an imaginary binary point just to the right of the sign bit (MSB), and therefore lie in the range from -1 to $+1$ (actually, $+1-2^{-24}$).

3.2.3.1. The storage array

The two port random access memory storage array is organised as 40 words of 25 bits each. The two 6-bit address fields of the instruction code allow two locations in RAM to be independently accessed. One location serves as a source only, and is read from the A port and passed to the ALU via the binary shifter. The other location is accessed through the B port, and acts as both the source for the second ALU input, and the destination for the ALU result.

Sixteen of the unused RAM addresses are used to implement a constant array. These four-bit constants, shown in table 3.1, are actually derived from the least significant four bits of the A address. The four bits are placed in the four most significant bits of the data field, with the remaining bits being set to zeroes. This provides an array of constants from -1 to $+\frac{7}{8}$ in $\frac{1}{8}$ increments. Other constants can be formed by passing these values through the scaler and accumulating

Table 3.1 Constant codes

Constant mnemonic	Unscaled value		Constant mnemonic	Unscaled value	
KP0	**.0**	**0.000**	**KM1**	**− .125**	**1.111**
KP1	**+.125**	**0.001**	**KM2**	**− .250**	**1.110**
KP2	**+.250**	**0.010**	**KM3**	**− .375**	**1.101**
KP3	**+.375**	**0.011**	**KM4**	**− .500**	**1.100**
KP4	**+.500**	**0.100**	**KM5**	**− .625**	**1.011**
KP5	**+.625**	**0.101**	**KM6**	**− .750**	**1.010**
KP6	**+.750**	**0.110**	**KM7**	**− .875**	**1.001**
KP7	**+.875**	**0.111**	**KM8**	**−1.0**	**1.000**

the result. Arbitrary constants of up to 25-bit precision can be formed in this way.

The 9-bit DAR can be used as a source or destination opcrand and occupies the nine most significant bits of a word which has the other bits set to all ones (a zero and all ones for the 2921). This corrects for an inherent A/D converter offset. The DAR is connected directly to thc D/A convcrtcr, which continuously converts the value in the DAR to an analog level. The DAR is also used as a successive approximation register during A/D conversions, and for conditional instructions.

3.2.3.2 Binary scaler

The scaler is an arithmetic barrel shifter which can be used to scale data read from the A port of the RAM before it is operated on by the ALU. Values read from the A port can be shifted up to two locations to the left, or thirteen locations to the right. Zeroes fill in from the right during left shifts. The sign bit is extended into the left-most bit positions during right shifts, preserving the correct sign of the shifted value.

3.2.3.3. ALU

The ALU calculates a 25-bit result from its A and B operands according to the 3-bit operation code, and writes this result to the B location of RAM. Operations executed by the ALU include addition, subtraction, AND, absolute value, and exclusive or. The instruction set is described further in section 3.3.

Although the digital processing unit uses 25-bit arithmetic, the ALU actually uses 28-bit precision by extending the sign bit of numbers to be processed to the left. This allows full scale numbers which are then left shifted to be processed by the ALU. The result will be correct unless it exceeds the 25-bit capacity of the remainder of the digital processing unit, in which case an overflow will occur.

In order to prevent instabilities in signal processing algorithms, and to make easier implementations of some non-linear functions, such as limiters, the 2920 ALU contains overflow saturation logic. This logic, which is normally enabled, preserves the sign of the result of an ALU operation which has overflowed, replacing the computed value with either + or − full scale. This logic can be disabled under software control, in which case the 28-bit ALU result is simply truncated to 25 bits.

Overflows are indicated by the active low, open drain, $\overline{\text{OF}}$ pin. The overflow is indicated the cycle following the instruction during which it occurred. The $\overline{\text{OF}}$ output can drive one TTL load with a suitable pull-up resistor. Disabling the overflow saturation logic disables the overflow output.

Some 2920 instructions can be executed conditionally, based on the state of selected bits in the DAR. Conditional instructions are described in section 3.3.

3.2.4 The Analog Section

A block diagram of the analog section of the 2920 is shown in fig. 3.2. This section includes a four input multiplexer, input and output sample-and-holds, a successive approximation A/D converter, a D/A converter, and an output demultiplexer. The DAR is also shown

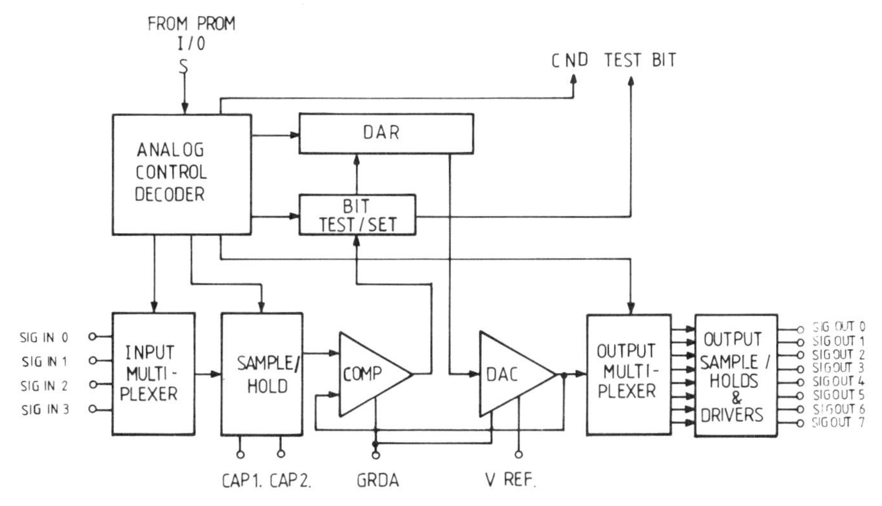

Figure 3.2 Analogue section block diagram

since all data flow to or from the analog section will involve this register. All operations of this section are controlled by the 5-bit analog instruction field.

The four analog inputs share a common sample-and-hold circuit. The sample-and-hold capacitor for this circuit is external on the 2920, and its value will affect the time constant of the sampling circuit. It will also affect offset due to charge sharing in the sampling switches. A value in the range from 100 pF to 1000 pF is recommended. A larger value will result in a smaller offset, but the acquisition time will increase. A value of approximately 500 pF will yield an offset of less then $-\frac{1}{2}$ LSB. The sample-and-hold capacitor is internal on the 2921.

The value stored on the sample-and-hold is converted to a digital value using a binary search successive approximation technique, which is under program control. The value in the DAR, which is normally cleared before beginning the conversion, is converted to an analog level by the D/A converter and compared to the input value. The output of the comparator is used to determine the current DAR bit, and the next lower bit is set to a one. The process continues until all bits have been converted. Anywhere from one to nine bits can be converted under program control.

An output operation uses the same D/A converter as is used for the A/D conversion. The output of the D/A converter is routed to the appropriate sample-and-hold circuit under program control. Outputs are held until updated during the next sample period.

For applications which require logic level outputs, two inputs, M1 and M2, can be utilised to choose either four or eight of the analog outputs to be placed in the logic mode. Table 3.2 lists the possible combinations. In this mode, the outputs become open drain, and a CMOS or TTL gate can be driven if a suitable pull-up is used.

The D/A converter requires an external positive voltage reference of between 1 and 2 volts. The step size for the D/A converter is VREF/256. Again, numbers are internally represented as fractions between +1 and −1, regardless of the value used for VREF. Any noise on the VREF pin will be transferred directly to the input and

Table 3.2 Output mode selection

M1	M2	SIGOUT Pins
5V	**5V**	**0–7 Analog**
5V	**−5V**	**0–3 Analog 4–7 Digital**
−5V	**5V**	**0–3 Digital 4–7 Analog**
−5V	**−5V**	**0–7 Digital**

output signals through the D/A converter. Hence, VREF noise should be kept to a minimum.

3.3 2920 Instruction Set

As previously mentioned, the 24-bit 2920 instructions are divided into five fields, each of which controls a section of the device. The format used by the 2920 assembler for these instructions is shown below in table 3.3. This format allows simultaneous analog and digital operations. A summary of the instruction set is given in table 3.4.

The ALU instructions are ADD, SUB, LDA (load), XOR, AND, ABS (absolute value), ABA (absolute value and add), and LIM. The ABS and ABA instructions convert the A source operand to its absolute value before any calculations are performed. The LIM instruction sets the result to +1 or −1 depending on the sign of the source operand.

Two special instructions, ABA . . . CND() and XOR . . . CND(), are used to disable and enable, respectively, the overflow saturation logic. The operands used will be affected, so care should be taken. The constant mnemonics, KPn and KMn, can be used for source and destination to avoid affecting any data since unused RAM address space will be accessed. The 2920 assembler will issue a warning because of the destination address, but executable code will be generated.

The ADD, SUB, and LDA instructions can be executed conditionally. A bit in the DAR, selected by the analog field, is used as a test bit to control instruction execution. If this bit is a one, the instruction executes; if the selected bit is a zero, a NOP results. Instruction execution time is the same in either case, ensuring a constant sample rate. Any of the nine DAR bits can be selected by the instruction in the analog field (CNDS though CND0). The conditional ADD and LDA operations allow decisions to be made in the 2920 program flow. The conditional ADD is also used for multiplication.

The conditional subtract is a special case, and is used to efficiently handle division. In this case, the subtract executes if the carry from the previous operation was a one, while an add is performed if the

Table 3.3 Instruction format

ALU INSTRUCTION (3 BITS)	B ADDRESS (6 BITS)	A ADDRESS (6 BITS)	SHIFT CODE (4 BITS)	ANALOG INSTRUCTION (5 BITS)

Table 3.4 Instruction set and op codes

ALU Instructions

Mnemonics	Operations		
ADD	**$(Ax2^N) + B$**	**$\rightarrow B$**	
SUB	**$B - (Ax2^N)$**	**$\rightarrow B$**	
LDA	**$(Ax2^N) + 0$**	**$\rightarrow B$**	
XOR	**$(Ax2^N) \oplus B$**	**$\rightarrow B$**	
AND	**$(Ax2^N) \cdot B$**	**$\rightarrow B$**	
ABS	**$\vert(Ax2^N)\vert$**	**$\rightarrow B$**	
ABA	**$\vert(Ax2^N)\vert + B$**	**$\rightarrow B$**	
LIM	**Sign (A) $\rightarrow \pm$ F.S.**	**$\rightarrow B$**	
ADD CND ( )(1)	**$(Ax2^N) + B$**	**$\rightarrow B$**	**IFF DAR (K) = 1**
	B	**$\rightarrow B$**	**IFF DAR (K) = 0**
SUB CND ( )(2)	**$B - (Ax2^N)$**	**$\rightarrow B$ & CY $\rightarrow$ DAR (K)**	**IFF CYp = 1**
	$B + (Ax2^N)$	**$\rightarrow B$ & CY $\rightarrow$ DAR (K)**	**IFF CYp = 0**
LDA CND ( )	**$(Ax2^N)$**	**$\rightarrow B$**	**IFF DAR (K) = 1**
	B	**$\rightarrow B$**	**IFF DAR (K) = 0**
ABA CND ( )	**$(Ax2^N) + B$**	**$\rightarrow B$**	**Disables overflow saturation**
XOR CND ( )	**$(Ax2^N) \oplus B$**	**$\rightarrow B$**	**Disables overflow saturation**

Analog Instructions

IN (K)	**Signal sample from input channel K**
OUT (K)	**D/A to output channel K**
CVTS	**Determine sign bit**
CVT (K)	**Perform A/D on bit K**
EOP	**Program counter to zero; enables overflow saturation**
NOP	**No operation**
CND (K)	**Select bit K for conditional instructions**
CNDS	**Select sign bit for conditional instructions**

Shifter Instructions

Mnemonics	Operations	Scale Factors
R13	**Shift right 13-bits**	**2^{-13}**
R12	**Shift right 12-bits**	**2^{-12}**
⋮	⋮	⋮
R01	**Shift right 1-bit**	**2^{-1}**
R00	**No shift**	**1**
L01	**Shift left 1-bit**	**2**
L02	**Shift left 2-bits**	**4**

Notes

(1) **CND ( )** can be either **CND (K)** or **CNDS** testing amplitude bits or the sign bit of the **DAR** respectively

(2) For **SUB CNDS** operation **CY $\rightarrow$ DAR (S)**

previous carry was a zero. In either case, the present carry is stored in the bit location of the DAR specified by the CND instruction.

The analog instructions for the 2920 are IN(K), OUT(K), CVT(K), EOP, NOP, and CND(K), which was already discussed. The IN(K) instruction, K = 0 to 3, selects one of the four 2920 inputs. The OUT(K) instruction, K = 0 to 7, selects one of the eight outputs. The CVT(K) instruction, K = S to 0, is used to convert the selected bit of the DAR during A/D conversion. An EOP causes the 2920 program counter to reset to zero after the next three instructions are executed. Again, analog instructions execute simultaneously with digital instructions.

An example 2920 instruction is shown below. This instruction performs the arithmetic operation $X = Y/4 + X$. During execution of this instruction, input channel zero is being sampled. The 2920 assembler accepts symbols X and Y for RAM addresses and automatically assigns RAM locations to the symbols as they are encountered during program assembly. Further examples of various 2920 instructions can be found in the applications section.

ADD X,Y,R02,IN0

3.4 2920 Development Process

Software support available for the 2920 includes an assembler, a software simulator, and the SPAS20 applications software. The assembler generates machine code from assembly language programs for use by the simulator, or for programming the 2920. The simulator allows the user to test and debug 2920 programs. Program execution can be traced, breakpoints set, program and data memory altered, and input signals specified. The SPAS20 Signal Processing Applications Software/Compiler is an interactive design tool for developing software for the 2920. It can generate optimised 2920 code for the functions developed using its interactive filter design and curve-fitting capabilities. These software support tools run on the Intellect Microcomputer Development Systems.

A system design kit, the SDK–2920, is also available for programming and evaluation of the 2920. It includes a one line assembler, an editor, a programming socket, upload and download capabilities through an RS232 port, and an applications section with breadboarding area for operating a programmed 2920.

It is convenient to approach a design using the 2920 by considering the functional blocks which can be implemented. Once a detailed block diagram of the system to be realised is developed, the individual functional blocks can be coded into 2920 code. Models exist for many commonly used functions. Several of these will be discussed below.

3.5 Arithmetic Building Blocks

Among the most essential building blocks, utilised by more complex functional blocks, are multiplication and division by both variables and constants. These simple building blocks are described in this section.

3.5.1. Multiplication by a Constant

In general, the 2920 instruction set is optimised to implement operations of the form

$$Y = Y + (C*X)$$

where C is an arbitrary constant. By taking advantage of the binary scaler in the ALU, such operations can be efficiently implemented by the 2920.

Consider the equation above. The constant C can be written as sums and differences of positive and negative powers of two. Assume that $C = 1.9375$. In binary, C would be 1.1111. This can also be represented as

$$C = 2^1 - 2^{-4}$$

The code which generates this constant would be

```
ADD Y,X,L01
SUB Y,X,R04
```

Note the direct correspondence between the 2920 code and the sum of powers of two representation.

An algorithm can be described which can be used to derive an efficient representation for a constant as a sum of powers of two.

Assume that the constant $C = 0.773$ must be represented with a tolerance of ± 0.001. The first step is to choose a power of two as a first estimate, V, of this value. In this case 2^0 is the closest, leaving an error $E = C - V = -0.227$. The power of two which is closest to this error is $-2^{-2} = -0.25$, and this becomes the next term of the expression. The new error is $E = C - V = 0.773 - 0.75 = 0.023$.

The next term should then be 2^{-5}, which leaves an error of -0.00825. A final term of -2^{-7} will leave an error of 0.0004375. The result is

$$V = 0.7734375 = 2^{0} - 2^{-2} + 2^{-5} - 2^{-7}\ (0.1100011).$$

In 2920 code

```
ADD Y,X
SUB Y,X,R02
ADD Y,X,R05
SUB Y,X,R07
```

Equations of the form $Y = C*X$ can be coded using the same techniques. In this case, the first instruction is LDA rather than ADD.

This algorithm produces efficient, though not always optimal 2920 code. Sometimes a more efficient representation can be found by observing the binary pattern of the constant. Consider the equation $Y = C*X$ using the value of C from the above example. The pattern 011 is repeated in the binary representation for C. A more efficient way to code this multiply would then be

```
LDA Y,X
SUB Y,X,R02
ADD Y,Y,R05   ;Y = Y(1+2^-5)
```

In some cases it is desirable to avoid intermediate results which are greater than the final result. In the case above, the code could be re-written as

```
LDA Y,X,R01
ADD Y,X,R02
ADD Y,Y,R05
```

with no penalty in the number of instructions required.

Sometimes it is necessary to implement equations of the form $X = C*X$. This can occur when memory space is becoming limited. In this case, C is represented as a product of factors of the form 1 ± 2^{n}. An algorithm to determine an efficient representation for C in this case is similar to one described above. Assume $C = 0.937$. A factor of the form 1 ± 2^{n} is chosen which is closest to the value of C, in this case, $1 - 2^{-4}$. This is divided into C, with a result of 0.99947. The factor which most closely equals this new value is $1 - 2^{-11}$. When this is divided into 0.99947 the result is 0.99995. The objective is to get this result to be close to one. The 2920 code at this point is

```
SUB X,X,R04
SUB X,X,R11   ; X=X(1-2^-4) (1-2^-11)=X(0.937003)
```

The number of 2920 instructions required for a multiplication by a constant sequence depends upon the value of *C*. However, such sequences are usually shorter than the corresponding multiplication by a variable procedure, which is described below. Note that because the constant never actually needs to be stored in the 2920 RAM, values larger than one can be implemented, as in the first example above. The SPAS20 compiler will generate optimised 2920 code for all of the equation types described above with user selectable error bounds.

3.5.2 Multiplication by a Variable

Multiplication by a variable is done using a shift and add algorithm. The conditional ADD instruction is used, conditioned on the bits of the multiplier. Program 3.1 shows the code for a complete four quadrant multiply.

The multiplier, MPLR, is loaded in the DAR so that the steps of the shift and add algorithm can be conditionally performed based on its value. For this reason, it is most efficient to limit the multiplier to nine bits. The multiplicand, MCND, is shifted and conditionally added to the partial product, PRO, based on the multiplier bits which are set in the DAR. If a multiplier bit is zero, then a multiply by zero should take place for that step, and this is done by not performing the add. The eight magnitude bits of the multiplier are handled first.

If the multiplier is negative, then the partial product must be

```
SUB PRO,PRO                 ; CLEAR THE PRODUCT
LDA DAR,MPLR                ; MULTIPLIER TO DAR
ADD PRO,MCND,R01,CND7       ;
ADD PRO,MCND,R02,CND6       ;
ADD PRO,MCND,R03,CND5       ;        THIS
ADD PRO,MCND,R04,CND4       ;       IS THE
ADD PRO,MCND,R05,CND3       ;      MULTIPLY
ADD PRO,MCND,R06,CND2       ;       PROCESS
ADD PRO,MCND,R07,CND1       ;
ADD PRO,MCND,R08,CND0       ;
SUB MCND,MCND,L01           ;   THESE TWO SUPPLY
ADD PRO,MCND,CNDS           ;   THE CORRECT SIGN.
SUB MCND,MCND,L01           ; RESTORE MULTIPLICAND
```

Program 3.1 Four quadrant multiply

corrected by subtracting the multiplicand from it. The conditional subtract instruction cannot be used here (see section 3.3) so, the multiplicand is negated by subtracting twice the multiplicand from itself, and the result added to the partial product conditional on the sign bit of the multiplier. This completes the multiply algorithm. The last instruction restores the sign of the multiplicand, and is only needed if this value will be used later in the program. The last three lines of code are not necessary if the multiplier will always be a positive number.

3.5.3 Division by a Variable

Division by a constant can be performed by multiplying by the inverse of the constant. Division of a variable by another variable uses the conditional subtract instruction. The algorithm described below is intended for positive variables. If negative variables must be processed, the magnitude should be used, and the sign of the quotient restored afterwards.

Division of two positive numbers can be accomplished by performing a series of test subtractions of the divisor from the dividend, shifting the divisor to the right before each test. If the result of a test subtraction is negative, the divisor is added back to restore the partial quotient, and the next test subtraction performed. The restoration and following test subtraction can be implemented in a single step by adding the divisor, after shifting it right once, if the previous result was negative. This is what the 2920 does. The conditional subtract operation will execute an add if the carry from the previous instruction was zero. The carry is always the complement of the sign of the previous result.

Program 3.2 lists the 2920 code to implement a four quadrant divide. The procedure involves conditionally subtracting the divisor from the dividend and assembling the quotient in the DAR. The first two lines of code extract the magnitudes of the numbers to be processed. The dividend should be scaled to ensure that it will be smaller than the divisor. The third line of code sets the carry initially so that the fourth line will execute a subtract. This fourth line will set the sign bit of the DAR if the divisor is larger than the dividend, which it should be. (For SUB CNDS, CY will be placed in the sign bit of the DAR.) The following eight lines perform the division algorithm described in the paragraph above. The thirteenth line forces an overflow if the sign bit of the DAR is a zero, which is only the case if the divisor was less than the dividend. This causes the result to

```
ABS DV1,DVD,R01              ; THESE TWO EXTRACT THE
ABS DV2,DVR                  ; MAGNITUDES
SUB DAR,DAR                  ; CLEAR THE DAR
SUB DV1,DV2,CNDS             ;
SUB DV1,DV2,R01,CND7         ;
SUB DV1,DV2,R02,CND6         ;        THIS IS
SUB DV1,DV2,R03,CND5         ;      THE DIVIDE,
SUB DV1,DV2,R04,CND4         ;      PROGRESSING
SUB DV1,DV2,R05,CND3         ;        ONE BIT
SUB DV1,DV2,R06,CND2         ;       AT A TIME.
SUB DV1,DV2,R07,CND1         ;
SUB DV1,DV2,R08,CND0         ;
ADD DAR,KP4,L01              ; THIS FORCES OVERFLOW
LDA QUO,DVD,R13              ; THESE TWO ESTABLISH
XOR QUO,DVR,R13              ; THE CORRECT SIGN
XOR QUO,DAR                  ; TRANSFER RESULT TO OUTPUT
```

Program 3.2 Four quadrant divide

saturate, preserving the sign. If the sign bit of the DAR had been a one, the sign bit would have been cleared. The last three lines determine the true sign of the result based on the original signs of the two operands.

If the dividend will always be less than the divisor, the third and thirteenth instructions can be removed, and the fourth instruction changes to an unconditional subtract. Additional instructions can be eliminated if the variables will be positive.

If greater precision is needed, the partial quotient in the DAR should be saved, before forcing the overflow and restoring the sign. The DAR can then be cleared, and the conditional subtractions continued once the carry is restored. This can be accomplished by adding and then subtracting the divisor, appropriately shifted, from the partial remainder.

3.6 Oscillators

There are several ways to implement oscillators with the 2920, including the use of unstable second order filters. However, one of the simplest approaches is to implement a sawtooth relaxation oscillator. The oscillator output can then be shaped to realise the desired waveform.

3.6.1 Relaxation Oscillator

A simple relation oscillator can be implemented by subtracting a constant $K1$ from one of the 2920 registers once each pass through the program. When the result becomes negative a second constant $K2$, larger than the first, is added to this register. This generates a sequence of samples for a negative sloping sawtooth waveform with a frequency equal to $(K1/K2)\,f_s$, where f_s is the sample rate. The amplitude varies from 0 to $K2$.

Another technique for generating a sawtooth oscillator takes advantage of the normal two's complement overflow by disabling the overflow saturation logic. A register can be continuously incremented or decremented producing a sawtooth of either slope due to the characteristics of a two's complement overflow. This technique may result in fewer instructions in some cases, and does not use the DAR. The saturation logic may need to be re-enabled for functions which follow. The EOP instruction will automatically enable the overflow saturation logic.

If a different waveform is desired, either filtering or waveform modification can be used. However, filtering should be used with caution on a waveform which is not bandlimited, such as the sawtooth. Aliasing components of the sampled sawtooth may pass through the filter and show up as amplitude modulation on the desired output signal. By performing a nonlinear transformation of the oscillator output, the samples can be modified to correspond to that of a more bandlimited waveform.

3.6.2 Sinusoidal Oscillators

The 2920 can implement a piecewise linear approximation to a sinewave of arbitrary accuracy. A particularly simple approximation is a triangle wave which has been clipped at two thirds of its amplitude. The waveform can be easily generated from the output of the sawtooth oscillator. The sawtooth waveform is simply centred about zero, converted to a triangle waveform using the absolute value instruction, re-centred about zero, and then multiplied by three to produce the clipped waveform, taking advantage of the overflow saturation logic.

```
SUB OSC,KP1              ; K1=1/8
LDA DAR,OSC
ADD OSC,KP4,L01,CNDS     ; RESET OSCILLATOR. K2=1
LDA SIN,OSC              ; BEGIN WAVEFORM SHAPING
SUB SIN,KP4              ; CENTRE SAWTOOTH ABOUT 0
ABS SIN,SIN,L01          ; CREATE TRIANGLE WAVE, 0 TO + 1
SUB SIN,KP4              ; CENTRE ABOUT 0
ADD SIN,SIN,L01          ; MULTIPLY BY THREE
```

Program 3.3 A simple sine oscillator

```
ABA TEMP,TEMP,CNDS     ; DISABLE OVERFLOW LIMITING
ADD SAW,KP2            ;
ADD SAW,KP3,R02        ; SAWTOOTH AT 1400.1 HZ
ADD SAW,KP3,R06        ; SAMPLE RATE = 8 kHZ
ADD SAW,KP7,R11        ;
XOR,TEMP,TEMP,CNDS     ; ENABLE OVERFLOW LIMITING
ABS TRI,SAW            ; FORM TRIANGULAR WAVEFORM,
ADD TRI,KM4            ; CENTRED ABOUT ZERO
LDA TEMP,TRI,L01       ; CODE FOR
LDA SIN, TRI, R02      ; SINE WAVEFORM
SUB SIN,TRI,R05        ; USING A PIECEWISE
ADD TEMP,TEMP,R02      ; LINEAR APPROXIMATION
ADD SIN,TEMP,R02
SUB SIN,TEMP,R05
ADD TEMP,TRI,R00
ADD SIN,TEMP,R03
SUB SIN,TEMP,R06
ADD TEMP,TRI,L01
ADD SIN,TEMP,R04
ADD SIN,TEMP,R09
LDA SIN,SIN,L01
```

Program 3.4 A sinusoidal oscillator with maximum absolute error of 0.009

The code to implement such an approximation to a sine wave at $\frac{1}{8}$ the sampling rate is shown in program 3.3.

This particular transformation has relatively low harmonic content when compared to the sawtooth. There are no even harmonics, or harmonics which are divisible by three. The existing odd harmonics fall off as $1/n^2$, with the fifth harmonic 28 dB down when compared to the fundamental, and the seventh 34 dB down.

If a more accurate representation of a sinewave is needed, the code listed in program 3.4 will produce a sinewave approximation using piecewise linear techniques with a maximum absolute error of 0.009.

3.6.3 Frequency and Phase Control

A voltage controlled oscillator or frequency modulated oscillator can be realised by varying the value of $K1$ for the basic oscillator described above. For example, the controlling signal could be scaled and added to a value $K0$ to produce the value $K1$. The constant $K0$ would represent the centre or free running frequency of the oscillator.

It is also easy to phase shift the oscillator signal, or to produce a phase modulated oscillator. Since the sawtooth is a linear, monotonic waveform, the phase is directly proportional to the amplitude. Hence, a phase change can be affected by adding or subtracting a constant

```
SUB OSC1,KP1         ; REFERENCE OSCILLATOR
LDA DAR,OSC1
ADD OSC1,KP4,L01     ; RESET OSCILLATOR IF LESS THAN 0
LDA OSC2,OSC1        ; OSC2 WILL BE THE PHASE SHIFTED OSCILLATOR
SUB OSC2,KP2         ; SUBTRACT 1/4 (90 DEGREE PHASE SHIFT)
LDA DAR,OSC2
ADD OSC2,KP4,L01     ; RESET IF NEGATIVE
```

Program 3.5 Phase shifting an oscillator

from the sawtooth. The code listed in program 3.5 demonstrates how to generate two waveforms which are 90 degrees out of phase. The individual sawtooth waveforms can be converted to sinusoidal waveforms using the approximation techniques described above.

Care should be taken when generating frequency or phase modulated signals since these signals are not band limited. The sampled versions of these signals generated inside the 2920 may be distorted by aliasing components if the sidebands of the modulated signals have significant spectral components above half the sampling rate.

3.7 Non-Linear Functions

Functions such as absolute value (ABS) and limit (LIM), conditional arithmetic, logical operations, and overflow can be used to implement non-linear functions in the 2920. These non-linear operations may be used for such purposes as simulating analog non-linear functions, and implementing piecewise linear approximation used for waveshaping, as was done for the sinewave. Again, caution must be observed since non-linear operations often produce signals rich in harmonics, and the sampled versions of these signals may be distorted due to aliasing.

A common non-linear operation is rectification. The ABS instruction of the 2920 implements a full-wave rectification. The ABA instruction is useful when combining full-wave rectification with input to a filter, as during measurement of signal amplitude. A half-wave rectification can be accomplished by implementing the equation $Y = (X + |X|)/2$ as follows

```
LDA Y,X,R01   ; Y=X/2
ABA Y,X,R01   ; Y=X/2+|X|/2
```

A hard limiter function can be implemented in the 2920 using the LIM instruction. This can be useful for producing a logic output from

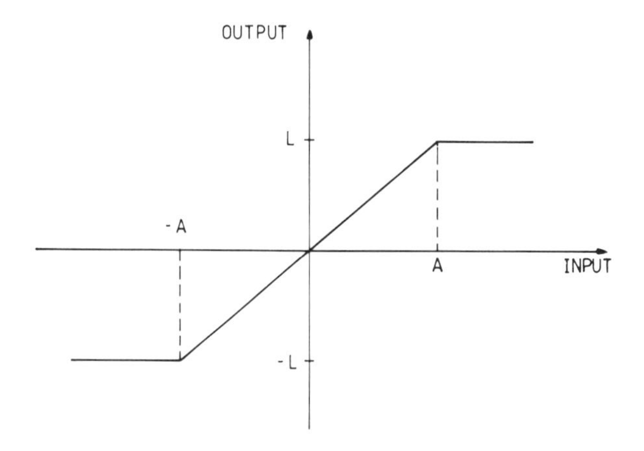

Figure 3.3 Limiter transfer characteristic

the 2920 to indicate when a threshold has been exceeded. The threshold is subtracted from the signal to be tested prior to the LIM instruction.

A limiter of the form shown in fig. 3.3 can be implemented in the 2920 using the overflow saturation logic. This is accomplished by multiplying (shifting) the input signal such that it will saturate when it exceeds the value A. The level L is achieved by further scaling of the result. The code below implements a limiter with $A = 0.1$ and $L = 0.75$.

```
LDA Y,X,L01  ; Y=2*X
ADD Y,Y,L02  ; Y=5*2*X=10*X
SUB Y,Y,R02  ; Y=Y-Y/4=0.75*Y
```

The technique just described suggests some interesting possibilities. If several transfer characteristics of the form of fig. 3.3 are generated with different slopes, scaled, and added together, complex functions can be approximated. This is exactly the type of piecewise linear technique use to produce the sinewave approximation generated by the code of program 3.4. A wide variety of functions can be produced in this way. The SPAS20 software package can generate piecewise linear approximations to arbitrary transfer functions (with some restrictions) and generate the 2920 code to implement them within user specified error limits.

3.8 Input and Output

The 2920 analog input and output functions are entirely under software control. The instruction sequences for these functions are programmed into the analog field of the instruction, and execute concurrently with the digital instruction, improving throughput.

A typical input sequence for the 2920 is shown in program 3.6. A sequence of IN(K) instructions is used to acquire a sample of the signal present on input channel K. The sampling switch is closed during this entire sequence. The number of IN(K) instructions needed is dependent upon the instruction cycle time and the size of the sample-and-hold capacitor. The sequence shown is for a 2920

```
SUB DAR,DAR,IN0
IN0
IN0
IN0
IN0
IN0
IN0
IN0
NOP
CVTS
NOP
NOP
CVT7
NOP
NOP
CVT6
NOP
NOP
CVT5
NOP
NOP
CVT4
NOP
NOP
CVT3
NOP
NOP
CVT2
NOP
NOP
CVT1
NOP
NOP
CVT0
```

Program 3.6 Typical 2920 input sequence

operating with a 600 ns instruction cycle and using a 500 pF sample-and-hold capacitor. The 2921, with its on-chip capacitor, would only require two IN(K) instructions when operating at the same clock rate. The analog NOP following the input sequence allows any transients to settle before the conversion process begins.

The DAR is cleared prior to beginning the A/D conversion. This is usually done during the first IN(K) instruction to allow the D/A converter to stabilise before conversion begins. The output of a comparator will adjust the selected bit in the DAR during the respective CVT(K) instruction. The next lower bit in the DAR is then set, and the process is repeated. One or two analog NOP instructions are placed between CVT(K) instructions to allow the D/A to settle after the DAR has been changed.

Any number of bits up to nine can be converted. Nine bits provide 54 dB of dynamic range (approximately 6 dB per bit). This dynamic range may not be needed to process some signals. Therefore a shorter conversion sequence can be used, saving analog instructions.

Although the 2920 was designed to operate directly on analog signals and hence has its I/O capabilities optimised for this task, it is possible to perform digital I/O.

A logic input can be performed with a one-bit conversion. A decision threshold can be set by loading the threshold value in the DAR prior to the conversion sequence. For example, assume that the signal present on input 2 can have two voltage levels, 0 V and 1 V. Further, assume that VREF = 1 V. One half would then be a suitable decision threshold, and this threshold would be loaded into the DAR. Converting bit 7 will result in that bit being cleared if the signal level is below one half. Otherwise that bit will be set. A conditional operation can then be executed based on bit 7. Fewer input instructions are required here than for the nine-bit conversion. The logic input can be used to control some function inside the 2920. This technique could also be used to implement a serial input to pass parameters to the 2920.

Program 3.7 shows a method which can be used if logic inputs are being performed on more than one input channel. In this case, the DAR is loaded with the threshold just once, and the input data packed into the least significant bits. The small changes in the threshold value stored in the DAR which may occur as the inputs are converted will be negligible compared to the signal level swing.

A typical output sequence is shown in program 3.8. The value to be output is placed in the DAR. A sequence of analog NOP instructions allows the D/A converter to settle. The OUT(K) instructions select the

```
LDA DAR,KP4,IN0
IN0
IN0
NOP
CVT1
IN1
IN1
IN1
NOP
CVT0
```

Program 3.7 Inputting logic values from two input channels

```
LDA DAR,OUTPUT
NOP
NOP
OUT0
OUT0
OUT0
OUT0
OUT0
OUT0
```

Program 3.8 Typical 2920 output sequence

desired output, and charge the corresponding sample-and-hold circuit. The output is held until it is updated on the next program pass. The sequence shown is for a 2920 operating with a 600 ns instruction cycle.

A logic output may be simulated using the analog outputs, or some of the 2920 outputs can be placed in the digital output mode, as described in section 3.2. An internal threshold of 1.5 V must be exceeded to produce a high level in this mode. Therefore, VREF must be greater than 1.5 V. A convenient way to assure that the proper output levels are generated is to use the LIM instruction to load the output value into the DAR. The overflow pin ($\overline{\text{OF}}$) may also be used as a digital output. Overflows can be forced based on the digital data to be output, and then clocked into a shift register with $\overline{\text{CCLK}}$.

The 2920 or 2921 data sheet should be consulted for current data on acquisition and settling times, and the appropriate input and output sequences for the particular application determined from this data.

3.9 Implementing Filters with the 2920

The 2920 can implement both recursive and non-recursive filter structures. However, because they provide a faster roll-off for a given filter order, recursive structures are more suited to the resources of the 2920. Therefore, recursive filters will be discussed.

3.9.1 Some Simple Filter Structures

A block diagram of a second order recursive filter is shown in fig. 3.4. The coefficients will be constants in the time invariant case. To implement this filter in the 2920 requires a delay line, and two constant multiplies of the type $Y = Y + C*X$ discussed in section 3.5.1.

There are several methods available for determining the coefficients to obtain a desired filter characteristic. One method which works well and is simple to implement is the matched-z transformation. This method provides a one-to-one mapping between poles/zeros in the s-plane and poles/zeros in the z-plane. The equations needed to perform this mapping for a complex pole are

$$B_1 = 2e^{-\sigma T} \cos wT \qquad B_2 = -e^{-2\sigma T}$$

where σ = real part of pole
w = imaginary part of pole
T = sample period

$$\text{maximum gain} = \frac{G}{(1 + B_2)(1 + [B_1^2/4B_2])^{1/2}}$$

Once the coefficients are determined, the code to implement the

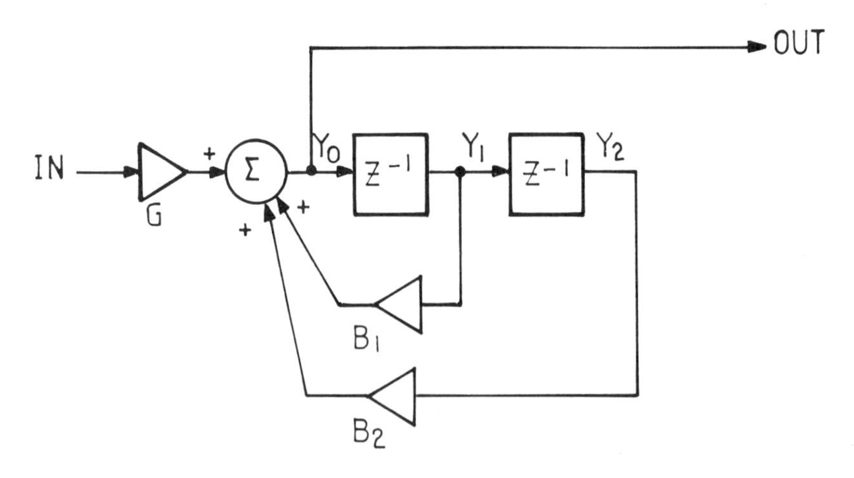

Figure 3.4 Digital filter implementation of a complex pole pair

```
LDA OUT2_P0,OUT1_P0,R00      ; IMPLEMENT DELAY LINE
LDA OUT1_P0,OUT0_P0,R00
LDA OUT0_P0,OUT2_P0,R04      ; BEGIN COEFFICIENT
SUB OUT0_P0,OUT2_P0,R01      ; CALCULATION
SUB OUT0_P0,OUT0_P0,R04      ; B2=-0.01101001
ADD OUT0_P0,OUT1_P0,R00
ADD OUT0_P0,OUT1_P0,R03
ADD OUT0_P0,OUT1_P0,R05      ; B1=1.00101
ADD OUT0_P0,IN0_P0,R03       ; ADD INPUT, G=0.125

; POLE LOCATION OF CORRESPONDING ANALOG FILTER (MATCHED-Z
; TRANSFORM) IS -567.36633, 566.30499 (REAL, IMAGINARY) IF SAMPLE
; RATE FOR THIS FILTER IS 8 kHz
```

Program 3.9 2920 code for a complex pole pair

required multiplies can be determined using the techniques described in section 3.5.1.

Program 3.9 shows 2920 code which implements a simple second order recursive filter. The filter in this example simulates a second order Butterworth filter which has a cut off at approximately $f_s/10$. The first two instructions implement the two stage tapped delay line shown in fig. 3.4. The filter taps are 2920 register locations (RAM). The delay line is implemented by shifting the values in these registers once each program pass, between consecutive executions of the coefficient multiplies. The value from delay element one is shifted to delay element two, and the value which was calculated as the filter output the previous sample period becomes the new output from the first delay element. A new filter output is then calculated. The filter output can actually be taken from any of the taps, as there will only be a linear phase difference between them.

The delay line tap values are multiplied by the filter coefficients and summed together along with a weighted version of the input. The input must be weighted by a factor G to prevent overflows since the filter will have a gain associated with it.

The filtered signal, contained in the RAM location OUT0_P0, should not be modified prior to being shifted to location OUT1_P0. Some care should be taken when ordering the instructions to avoid situations which may lead to intermediate overflows.

A simple first order filter section simply deletes one of the delay elements. The single coefficient can be calculated as

$$B = e^{-\sigma T}$$

The maximum gain, $G/(1 - B)$, occurs at dc.

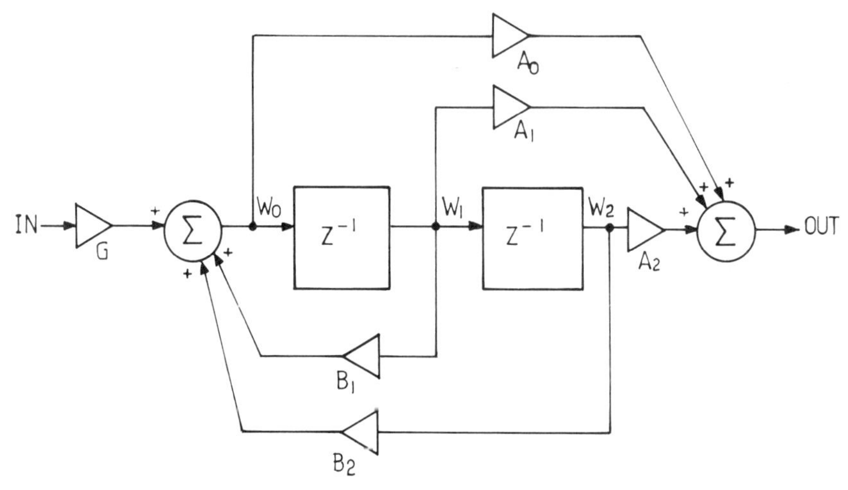

Figure 3.5 Digital filter implementation of a complex pole and zero pair

By providing weighted feed-forward paths for the delay line and summing these together, zeros can be implemented. Fig. 3.5 shows how a complex zero can be added to the complex pole of fig. 3.4. The formulae for calculating the coefficients for complex zeros using the matched-z transformation are

$$A_0 = \text{constant}$$
$$A_1 = 2A_0 e^{-\sigma T} \cos wT$$
$$A_2 = A_0 e^{-2\sigma T}$$

and for real zeros,

$$A_0 = \text{constant}$$
$$A_1 = A_0 e^{-\sigma T}$$

A real zero does not use the second delay element. The maximum possible gain for a zero is the sum of the absolute values of all the coefficients.

A zero is usually associated with a pole. By combining a pole and zero as shown in fig. 3.5, the number of delay elements (RAM locations) required is reduced since the tapped delay line is shared between the pole and zero. The gain for the overall quadratic section of fig. 3.5 is not the product of the individual maximum gains of the pole and zero pairs because these individual maximum gains do not occur at the same frequency.

Optimised code for filter sections of the form described above can be generated by the SPAS20 software package given the locations of the poles and/or zeros to be implemented. The matched-z transformation is used if the s-plane pole/zero locations are entered. Frequency responses corresponding to entered pole and zero pairs can also be plotted.

3.9.2 Cascading Filter Sections

To implement more complex filters, sections such as those described above can be cascaded or connected in parallel. An example will serve to illustrate some of the considerations when cascading filter sections. Assume that a six pole, three zero filter is to be implemented with a gain characteristic as shown in fig. 3.6. This plot was generated by the SPAS20 applications software package. The corresponding pole/zero locations are shown. These locations may have been determined by several means, including interactive design using the SPAS20 applications software, direct design in the z-domain, or s-plane design with a suitable transformation to arrive at the z-plane pole/zero locations.

The structure corresponding to the digital implementation of this

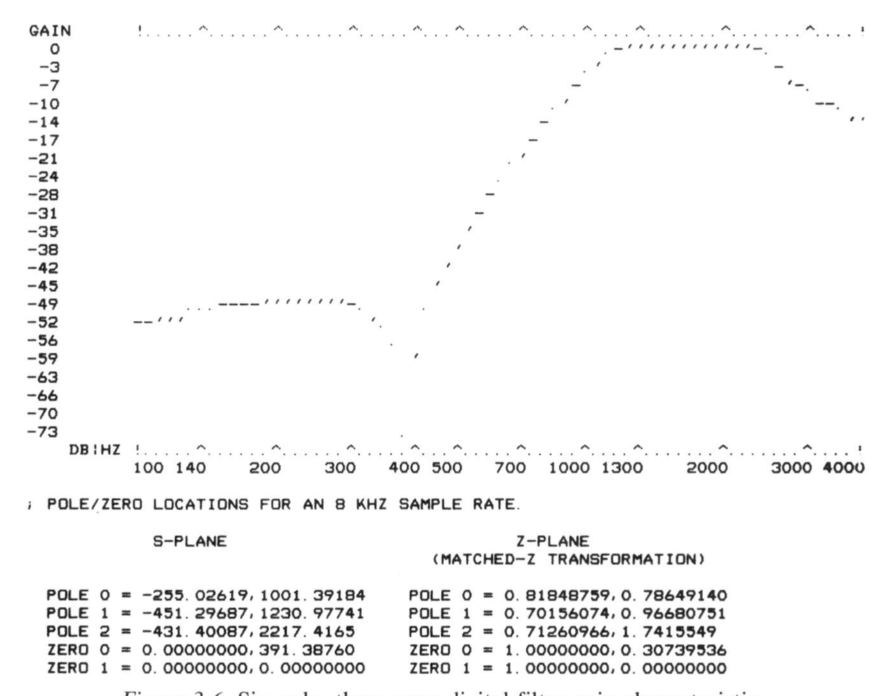

Figure 3.6 Six pole, three zero digital filter gain characteristic

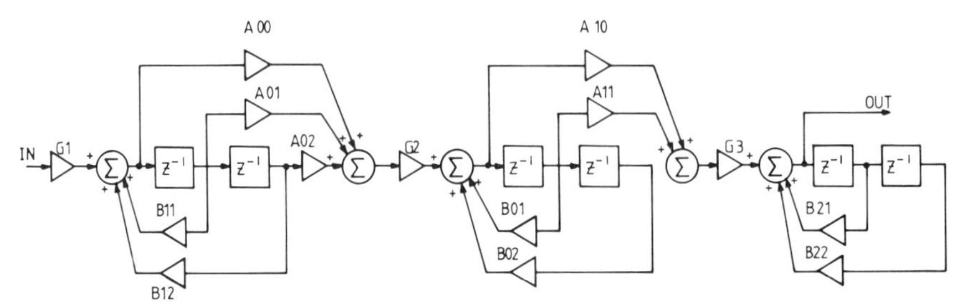

Figure 3.7 Digital implementation of six pole, three zero filter

filter is shown in fig. 3.7. Pole and zero structures are combined, and the resulting quadratic sections cascaded. The 2920 code for the individual filter sections is shown in program 3.10. This is how the code will look after being generated by the SPAS20 compiler (comments generated by the compiler have been removed). These individual sections of code must be connected together to realise the filter of fig. 3.6. It is most efficient from the standpoint of instructions required to pair the individual pole and zero sections to form quadratic sections. These sections can then be arranged to form the overall filter. The next step is to determine the scaling of the filter input required to prevent overflows.

The maximum overall gain for this filter is about 4.4. Scaling the filter input by 1/4.4 will provide an overall gain of one. However, since intermediate calculations may result in an intermediate gain in excess of the overall filter gain, the input is usually scaled to the overall gain which is less than one. A value between 0.25 and 0.5 is usually sufficient. It is code efficient to choose a power of two for the scaling factor, so that it can be implemented as a simple right shift. It is usually best to distribute the scaling among the individual filter sections. This helps avoid shifting the input too far right, which, when combined with the right shifts which implement the filter coefficients, may cause truncation errors. If the filter input variable has nine bit precision, it can be sufficient to avoid overflows, but not so excessive as to adversely affect filter accuracy.

Scaling is most easily done at the input to each of the quadratic sections. To know how best to distribute the scaling, the contribution to the overall filter gain of each of the quadratic sections should be determined. These calculations will also uncover any possible problems with intermediate overflows which might be caused by a filter stage with particularly high gain. A good way to proceed is to remove the individual poles and zeros from the filter, one at a time, starting with the last one to be implemented. The maximum filter gain

```
LDA OUT2_P1,OUT1_P1,R00      ; POLE 1
LDA OUT1_P1,OUT0_P1,R00
SUB OUT0_P1,OUT1_P1,R02
ADD OUT0_P1,OUT0_P1,R04
SUB OUT0_P1,OUT2_P1,R01
ADD OUT0_P1,OUT2_P1,R07
ADD OUT0_P1,IN0_P1,R00

LDA OUT0_Z0,IN2_Z0,R00       ; ZERO 0
SUB OUT0_Z0,IN1_Z0,L01
ADD OUT0_Z0,IN1_Z0,R04
ADD OUT0_Z0,IN1_Z0,R05
ADD OUT0_Z0,IN0_Z0,R00

LDA OUT2_P0,OUT1_P0,R00      ; POLE 0
LDA OUT1_P0,OUT0_P0,R00
LDA OUT0_P0,OUT2_P0,R02
SUB OUT0_P0,OUT2_P0,R06
SUB OUT0_P0,OUT2_P0,R00
SUB OUT0_P0,OUT0_P0,R03
ADD OUT0_P0,OUT1_P0,R00
ADD OUT0_P0,OUT1_P0,R03
ADD OUT0_P0,OUT1_P0,R05
ADD OUT0_P0,IN0_P0,R00

LDA OUT0_Z1,IN0_Z1,R00       ; ZERO 1
SUB OUT0_Z1,IN1_Z1,R00

LDA OUT2_P2,OUT1_P2,R00
LDA OUT1_P2,OUT0_P2,R00
LDA OUT0_P2,IN0_P2,R00
SUB OUT0_P2,OUT2_P2,R01
SUB OUT0_P2,OUT2_P2,R07
SUB OUT0_P2,OUT1_P2,R02
ADD OUT0_P2,OUT1_P2,R07
```

Program 3.10 2920 code for the individual poles and zeros of the six poles, three zero filter

is calculated at each step. This provides information on the contribution of each section to the overall filter gain. The distribution of scaling can then be chosen. The SPAS20 software can be used to calculate the gains. An alternative approach is to calculate the maximum gain of each section using the formulae of section 3.9.1, and scale accordingly. The effects of the zeros can usually be ignored, and scaling based on the poles. However, since maximum gains of the various stages occur at different frequencies, more scaling than necessary may result.

When scaling for a quadratic section, the gain of the pole should be used as a guide since it is the first element to be implemented. The overall section gain will then usually be less than one due to attenuation from the zero. This can be taken into account when scaling for the following section.

In the example, the input to the first stage was scaled by 1/8. The resultant gain of the first section is then about 2.2/8 = 0.27. A total attenuation of at least 1/5.6 = 0.18 is required for the next stage. An additional right shift of the signal going into this stage will provide this with some room to spare for possible intermediate overflows. The gain after the first two stages is 0.13, which is sufficient for the entire filter

```
SUB DAR,DAR IN0                     ; CLEAR DAR FOR INPUT
LDA TMP1,OUT1_P1,R00,IN0            ; BEGIN POLE 1
LDA OUT1_P1,OUT0_P1,R00,IN0
SUB OUT0_P1,OUT1_P1,R02,IN0         ; SYMBOLS OF THE FORM OUT0_P1
ADD OUT0_P1,OUT0_P1,R04,IN0         ; WERE GENERATED BY THE SPAS20
SUB OUT0_P1,TMP1,R01,IN0            ; COMPILER.
ADD OUT0_P1,TMP1,R07,IN0
ADD OUT0_P1,SIGIN,R03,IN0           ; INPUT RECEIVED SIGNAL TO FILTER.
SUB TMP1,OUT1_P1,L01                ; BEGIN ZERO 0
ADD TMP1,OUT1_P1,R04,CVTS
ADD TMP1,OUT1_P1,R05
ADD TMP1,OUT0_P1,R00
LDA TMP2,OUT1_P0,R00,CVT7           ; BEGIN POLE 0
LDA OUT1_P0,OUT0_P0,R00
LDA OUT0_P0,TMP2,R02
SUB OUT0_P0,TMP2,R06,CVT6
SUB OUT0_P0,TMP2,R00
SUB OUT0_P0,OUT0_P0,R03
ADD OUT0_P0,OUT1_P0,R00,CVT5
ADD OUT0_P0,OUT1_P0,R03
ADD OUT0_P0,OUT1_P0,R05
ADD OUT0_P0,TMP1,R01,CVT4
LDA TMP1,OUT0_P0,R00                ; BEGIN ZERO 1
SUB TMP1,OUT1_P0,R00
LDA TMP2,OUT1_P2,R00,CVT3           ; BEGIN POLE 2
LDA OUT1_P2,OUT0_P2,R00
LDA OUT0_P2,TMP1,R00
SUB OUT0_P2,TMP2,R01,CVT2
SUB OUT0_P2,TMP2,R07
SUB OUT0_P2,OUT1_P2,R02
ADD OUT0_P2,OUT1_P2,R07,CVT1        ; OUTPUT OF FILTER IN OUT0_P2
NOP
NOP
CVT0
LDA SIGIN,DAR                       ; STORE INPUT IN SIGIN
```

Program 3.11 2920 code for the filter after merging the poles and zeros

since the last stage actually reduces the overall gain. The overall gain is then 4.4/16 = 0.28. Some trial and error on the ordering of the poles and zeros may be necessary to achieve the optimal configuration for gain considerations.

The completed code for the filter is shown in program 3.11. To merge the poles and zeros, the outputs of the delay stages of the poles become the inputs of the zeros. For example, in the first quadratic section of the code of program 3.10, OUT0_P1 becomes IN0_Z0, and OUT1_P1 becomes IN1_Z0. An input sequence is also shown to illustrate how an input sample can be acquired while digital instructions are being executed.

Some adjustments of the 2920 code were made to reduce the number of instructions and RAM locations used by the filter. For example, the output of the last delay element of a quadratic section is not needed once the calculations for that section are complete. Therefore, the memory location which stores this value can be used again in following calculations. Such temporary RAM locations which are used in this example are denoted by TMP1 and TMP2. The RAM location used to accumulate the sum of the feed forward paths of the zero can also be used again once the value in it has been transferred to the following stage. Furthermore, in some cases the temporary RAM location used by a zero to accumulate the result can be the same one used by the associated complex pole to store the output of the second delay element. This was done for the first quadratic section in the example, and it resulted in the elimination of the instruction

```
LDA OUT0_Z0,OUT2_P1,R00
```

which became

```
LDA TMP1,TMP1,R00
```

after the variable substitution, and could therefore be eliminated.

3.9.3 Allpass Networks

In some applications it is necessary to have very little phase distortion of the signal being filtered. A linear phase filter can be constructed using FIR filter techniques, but for a filter response like that of fig. 3.6, many taps would be needed. It is often possible to achieve acceptable phase characteristics by adding an all pass network to an IIR filter to adjust its phase characteristics. This approach will usually require fewer 2920 instructions than a FIR filter implementation.

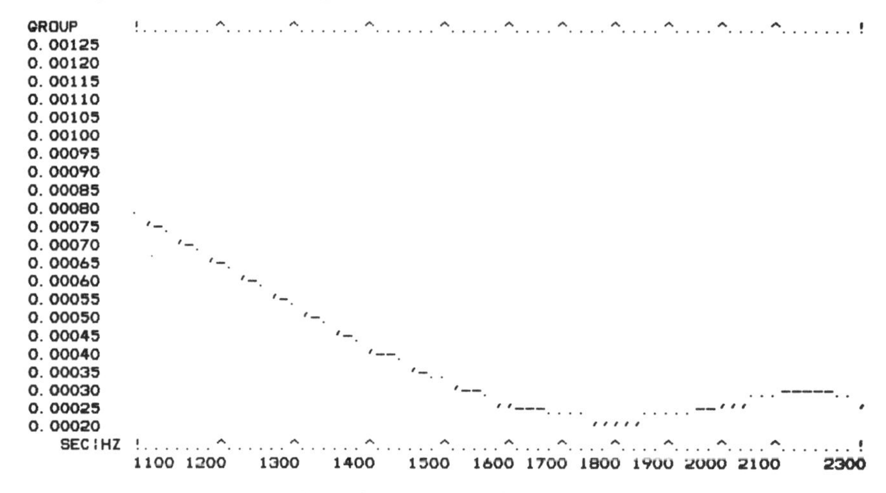

Figure 3.8 Filter group delay characteristic

Assume that the filter described above will be used in an application which requires minimal phase distortion in the frequency band from 1100 Hz to 2300 Hz. The group delay characteristic of this filter over the band of interest is shown in fig. 3.8. This characteristic can be flattened by passing the output signal of the filter through a network which will add more delay to the higher frequencies without affecting

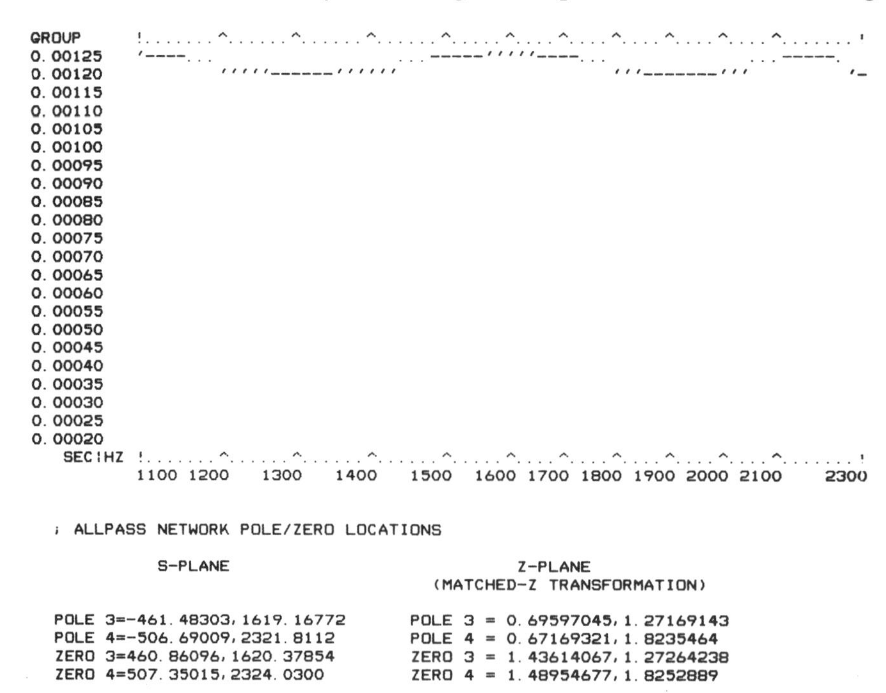

Figure 3.9 Group delay characteristic after equalisation

```
LDA TMP2,OUT1_P3,R00          ; BEGIN POLE 3
LDA OUT1_P3,OUT0_P3,R00
LDA OUT0_P3,OUT1_P3,R01
SUB OUT0_P3,OUT1_P3,R04
SUB OUT0_P3,OUT0_P3,R04
SUB OUT0_P3,TMP2,R01
ADD OUT0_P3,TMP2,R06
ADD OUT0_P3,OUT0_P2,R01       ; INPUT TO ALLPASS NETWORK
LDA TMP1,TMP2,R04             ; BEGIN ZERO 3
ADD TMP1,TMP2,L01
SUB TMP1,OUT1_P3,R00
ADD TMP1,OUT1_P3,R03
ADD TMP1,OUT1_P3,R05
ADD TMP1,OUT0_P3,R00
LDA TMP2,OUT1_P4,R00          ; BEGIN POLE 4
LDA OUT1_P4,OUT0_P4,R00
LDA OUT0_P4,TMP2,R04
SUB OUT0_P4,TMP2,R01
ADD OUT0_P4,OUT0_P4,R05
SUB OUT0_P4,OUT1_P4,R02
SUB OUT0_P4,OUT1_P4,R04
SUB OUT0_P4,OUT1_P4,R06
SUB OUT0_P4,OUT1_P4,R07
ADD OUT0_P4,TMP1,R01
LDA TMP1,TMP2,R02             ; BEGIN ZERO 4
SUB TMP1,TMP2,R05
ADD TMP1,TMP2,L01
ADD TMP1,OUT1_P4,R00
SUB TMP1,OUT1_P4,R02
ADD TMP1,OUT0_P4,R00          ; TMP1 IS OUTPUT OF ALLPASS NETWORK
```

Program 3.12 2920 code for the allpass network

the magnitude response of the filter. Such an allpass network, in the s-plane, consists of a pair of complex poles with a corresponding pair of complex zeros which are at the same frequency, but in the right-half plane. Using a fourth order allpass network, the delay differential in this example was reduced from 600 μs to 59 μs. The resulting group delay characteristic can be seen in fig. 3.9. The code to implement this network is shown in program 3.12.

3.9.4 A Particularly Simple One Pole Filter Implementation

The filter described below is particularly useful when a simple low-pass filter must be implemented and instruction locations are scarce. This filter requires just two instructions. It takes advantage of the fact that a constant multiply of the form $1 - 2^{-n}$ can be done with one instruction.

This, of course, limits the number of possible coefficients, and hence the possible filter cut-off frequencies. The code for such a one pole filter with a cut-off of 40.4 Hz at a sample rate of 8 kHz is shown below.

```
SUB LPF,LPF,R05   ; NEW LPF = OLD LPF(1-1/32)
ADD LPF,SIG,R05   ; ADD INPUT, MAX GAIN (DC)=1
```

Because the coefficient calculation requires just one instruction, the result of the multiply can be stored back to the same RAM location from which the source was read. This saves an instruction, the one which implements the delay line. The one sample period delay between filter calculations takes care of the delay. The input should be right shifted by the same amount used for the coefficient calculation in order to achieve unity gain.

3.9.5 Low Frequency Filters

The coefficients of very low frequency filters (when compared to the sample rate) can be difficult to implement due to the precision required. Even the 25-bit arithmetic of the 2920 may not be sufficient. Also, low frequency filters can have very high gain. A convenient way to get around these problems is to reduce the sample rate of the filter. This can be accomplished by using conditional instructions which are triggered by an oscillator running at a submultiple of the sample rate.

The code for a two pole low-pass filter implemented at 1/100 of the 2920 sample rate is shown in program 3.13 along with the oscillator

```
LDA OUT0_P0,OUT1_P0             ; RESTORE OUT0_P0
SUB OUT0_P0,OUT1_P0,R05         ; BEGIN FILTER CALCULATION
ADD OUT0_P0,OUT0_P0,R01
SUB OUT0_P0,OUT2_P0,R01
SUB OUT0_P0,OUT2_P0,R04
SUB OUT0_P0,OUT2_P0,R07
ADD OUT0_P0,IN0_P0,R05          ; TOTAL GAIN = 8.5/32 = 0.27
SUB TIMER,KP1,R05               ; BEGIN SAMPLING TIMER
LDA DAR,TIMER
LDA TIMER,KP3,CNDS              ; RESET TIMER TO
ADD TIMER,KP3,R05,CNDS          ; 99 TIMES KP1,R05
LDA OUT2_P0,OUT1_P0,CNDS        ; IMPLEMENT DELAY LINE ONLY
LDA OUT1_P0,OUT0_P0,CNDS        ; WHEN TIMER IS RESET
```

Program 3.13 2920 code for a filter operating at 1/100 of the sample rate used for the submultiple sampling. The filter is designed for the lower sample rate and has its cutoff at 1/16 of that rate in this example. The filter code is similar to that for any second order section, except that the instructions which implement the delay line are conditional, and only execute at the reduced sample rate, when the oscillator resets. The entire filter could be implemented conditionally, but more instructions would be required since no subtractions could be performed.

The key to implementing the submultiple sampling is to make sure the values in the delay line are not modified between operations. This is easy to do for the values OUT2_P0 and OUT1_P0, by using conditional loads, but can be more difficult for OUT0_P0, which is recalculated each pass through the 2920 program. By putting the delay line after the filter calculations, the new value of OUT0_P0 is saved in OUT1_P0 when the oscillator resets. The instruction right before the filter calculations restores the value of OUT0_P0. This instruction is only needed if the filter code uses the previous value of OUT0_P0, as in this example (second instruction). The value OUT1_P0 can be used as the output of the filter, making OUT0_P0 a temporary variable. Note that OUT2_P0 cannot be used as a temporary as was done in the complex filter example above.

An additional anti-aliasing filter may be needed for the low frequency filter depending on the application. This filter can be implemented digitally as long as suitable anti-aliasing precautions are taken for the digital anti-aliasing filter.

3.9.6 Time Variable Filters

Filters with variable coefficients can be implemented with the 2920 if the constant multiplies used in the previous filter examples are replaced by variable multiplies. This will allow control of the filter bandwidth or centre frequency by one of the 2920 inputs, which can be either analog or serial digital, or some other portion of the program.

For a single complex pole with a structure as shown in fig. 3.4, the following approximations may be used

$$b \cong (-f_s/2\pi)\log(-B_2) \qquad B_2 \cong -\exp(-2\pi b/f_s)$$

$$f_o \cong (f_s/2\pi)\cos^{-1}(B_1/2\sqrt{B_2}) \qquad B_1 \cong 2\exp(-\pi b/f_s)\cos(2\pi f_o/f_s)$$

where f_s is the sample rate, f_o is the centre frequency of the filter, and b is its bandwidth. Note that if only bandwidth is to be controlled while the centre frequency remains fixed, both coefficients must be varied.

More complex filters can be implemented by cascading such second order sections.

As an example, suppose that a filter with a centre frequency which can be varied between 500 Hz and 1000 Hz is to be implemented. This filter should have a bandwidth of 40 Hz ± 5 Hz. The sample rate is 8 kHz. Using the approximations above, the range of values for B_2 is

$$-0.9729 \leqq B_2 \leqq -0.9653$$

The value $B_2 = -0.96875 = -(1 - 1/32)$ can be implemented with two 2920 instructions, and provides a bandwidth of 40.4 Hz. Using this value for B_2, the range for B_1 is found to be

$$1.8187 \geqq B_1 \geqq 1.3919$$

The coefficient B_1 can be implemented as $B_1 = 1.0 \pm C$, where C is the control variable. This keeps the value for C less than 1, as it must be. Varying C from 0.8187 to 0.3919 will cause the filter to sweep over the desired range.

The gain of the filter will vary as C is changed. Using the maximum gain formula of section 3.9.1, the gain is seen to vary from 83.6 for $C = 0.8187$ to 45.3 for $C = 0.3919$, which is a 5.3 dB variation. If this

```
LDA Y2,Y1
LDA Y1,Y0
ADD Y0,Y2,R05
SUB Y0,Y2                  ; Y0=B2*Y2+Y1
LDA TMP,X,R06              ; BEGIN GAIN CORRECTED INPUT
SUB TMP,X,R09              ; TMP=-0.01365*X
ADD TMP,Y1                 ; TMP=Y1-0.01365*X
LDA DAR,C                  ; SET UP FOR MULTIPLY
ADD Y0,TMP,R01,CND7        ; MULTIPLY BY C
ADD Y0,TMP,R02,CND6
ADD Y0,TMP,R03,CND5
ADD Y0,TMP,R04,CND4
ADD Y0,TMP,R05,CND3
ADD Y0,TMP,R06,CND2
ADD Y0,TMP,R07,CND1
ADD Y0,TMP,R08,CND0        ; Y0=B2*Y2+(1+C)Y1-0.01365*C*X
                           ;   =B2*Y2+B1*Y1-0.01365*C*X
ADD Y0,X,R06
ADD Y0,X,R09
SUB Y0,X,R11               ; FINISH BY ADDING 0.01709*X
```

Program 3.14 2920 instructions for a filter with variable centre frequency

variation is not tolerable, a weighting of the input of the form

$$X(0.01709 - 0.01365C)$$

will prevent overflow and yield a gain variation of approximately 0.5 dB. The code for the variable filter and the gain adjustment is shown in program 3.14. By summing $Y1$ and $-0.01365*X$ into TMP, only one multiply by the variable C need be done.

If just a few different filter characteristics are desired, it may be more efficient to implement the desired filters separately and use conditional instructions, based on a logic input, to choose the desired filter output. The efficiency of this approach will depend on filter complexity.

3.10 Some Other Building Blocks

Several additional functions which demonstrate some useful programming techniques are described below.

3.10.1 Peak Detector

The peak detector described here is similar to an analog peak detector implemented with an RC network. It is 'charged' to the input signal level whenever that level exceeds the output. When the input level is below the output level, the stored value decays in an exponential fashion. The RC portion of the peak detector is simulated with a one pole digital filter whose time constant depends on the location of the pole. An example is given below.

```
LDA DAR,OUT         ; PUT OUTPUT IN DAR
SUB DAR,IN          ; DAR=OUT-IN
SUB OUT,OUT,R08     ; DECAY OF OUTPUT (ONE POLE FILTER)
LDA OUT,IN,CNDS     ; CHARGE FILTER IF IN>OUT
```

The one pole digital filter shown above has a coefficient of $B = 1 - 2^{-8}$. The time constant can be calculated as

$$\tau = -f_s \log B$$

With a sample rate (f_s) of 8 kHz, the time constant would be 32 ms for this example.

3.10.2 Level Detector With Hysteresis

A level detector is implemented by subtracting a threshold value from the input signal level, and using the sign of the result for making a decision. The average signal level can be determined by rectifying and low-pass filtering the signal. In some cases it is desirable to introduce some amount of hysteresis to the threshold detection to avoid dithering about the decision point. The code shown below implements a level detector which has 4.4 dB of hysteresis.

```
; SAMPLE RATE IS 8 KHZ

SUB  LPF,LPF,R06              ; 20 HZ LPF
ABA  LPF,SIG,R06              ; FULL WAVE RECTIFIER
LDA  DAR,PAST                 ; PUT PAST LEVEL IN DAR
LDA  PAST,LPF                 ; PUT NEW LEVEL IN PAST
SUB  PAST,KP3,R05             ; COMPARE TO -34.7 DB
                              ; THRESHOLD
ADD  PAST,KM4,R06,CNDS        ; SET THRESHOLD TO -30.3 DB IF
                              ; SIGNAL WAS NOT PRESENT THE
                              ; PREVIOUS SAMPLE PERIOD
```

The full-wave rectifier produces a dc level of $2/\pi = 0.6366$ for a sinusoidal input of amplitude one. It is this value which must be considered when calculating the threshold levels. Note that there is a delay of one sample period between the threshold comparison and the conditional operations based on it.

3.10.3 Zero Crossing Detectors

A zero crossing detector can conveniently be implemented using the XOR instruction. The past and present value of the signal are simply exclusive-ORed together and the zero crossing detect implemented with a conditional instruction based on the sign of the result. The sign bit will be set if the signs of the two compared samples differ. To detect a positive transition, the result of the XOR instruction should be ANDed with the previous sample, before any conditional operations are performed. For a negative transition, the current sample should be used. An alternative method for implementing a negative transition detector is shown below.

```
; ASSUME THE PRESENT INPUT SAMPLE IS IN THE DAR

LDA TEMP,DAR    ; SAVE PRESENT INPUT
LIM DAR,DAR     ; LIMIT THE PRESENT SAMPLE
SUB DAR,PAST    ; SUBTRACT THE PAST VALUE
LIM PAST,TEMP   ; LIMIT PRESENT VALUE AND SAVE
```

The sign bit of the DAR will only be set when the present sample is negative and the past sample is positive. This demonstrates how arithmetic instructions coupled with the 2920 overflow saturation logic can sometimes be used to implement logic functions. In a similar fashion, the ADD instruction can often be used to perform a logical OR.

3.11 Hardware Considerations

Because the primary interface of the 2920 is analog, standard analog layout techniques should be followed when using the 2920. In general, analog and digital signals should be separated where possible. Noise on VREF and GRDA should be kept below 4 mV to prevent A/D conversion errors. The GRDA and GRDD leads are not connected internally on the 2920, and must be connected outside the device. To reduce noise, this connection is usually made at the point where ground is brought onto the board, and a separate low impedance path to GRDA provided. Standard analog power supply bypass techniques should be followed.

The typical hardware configuration for the 2920 is pictured in fig. 3.10. The 2920 is shown with a 6.144 MHz crystal, which will provide an 8 kHz sample rate for a full length program. With the output configuration pin M1 and M2 connected as shown, SIGOUT4 through SIGOUT7 will be in the digital mode. These outputs can drive one TTL load with the pull-up resistors shown. The pull-up resistors on $\overline{\text{OF}}$ and $\overline{\text{CCLK}}$ are not required if these signals are unused. If a 2921 is used, the CAP1, and CAP2 pins can float. To reduce input crosstalk and avoid input signal loading during sampling, the 2920 SIGIN pins should be driven from low impedance sources. Typical impedance during sampling for the 2920 is around 1 kΩ. The 2920 and 2921 data sheets should be consulted for exact input and output characteristic.

A circuit for buffering and filtering the reference voltage is shown. This circuit will provide a low noise reference voltage at the VREF pin. A simpler VREF circuit can certainly be used, but care should be taken since any noise or variation of VREF will affect A/D and D/A

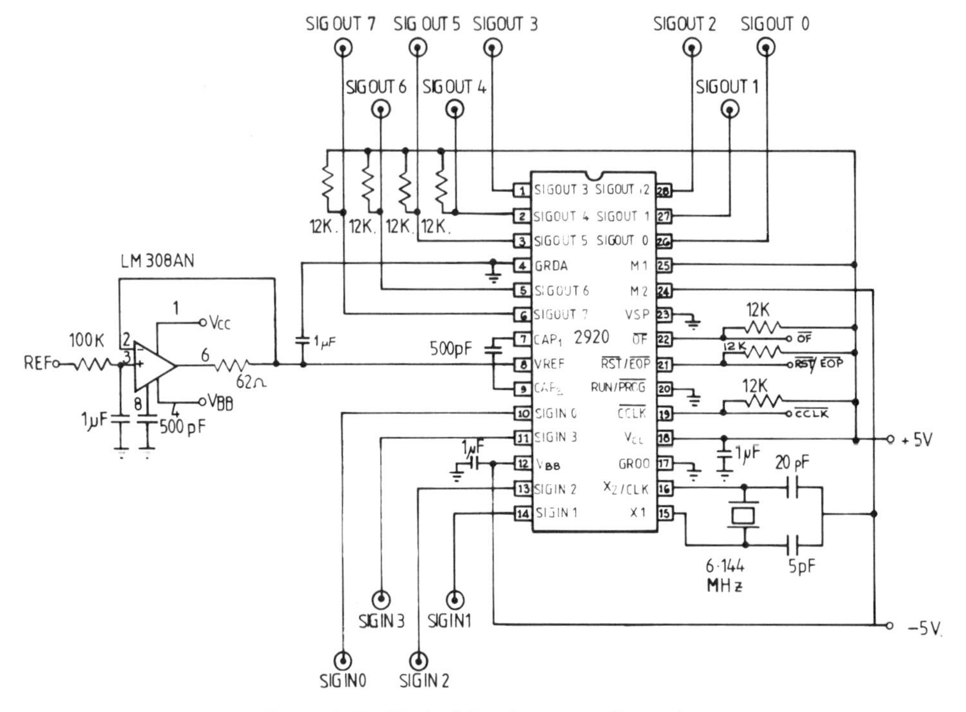

Figure 3.10 Typical hardware configuration

conversions. The reference voltage source is not shown, but can be generated by several means, including a resistive network to divide down the supply voltage, or a zener diode.

The 2920 can be driven from an external clock source connected to pin 16. The applied clock signal should oscillate between 0 V and −5 V. A possible circuit for driving pin 16 is given in figure 3.11. Driving the clock pins differentially will result in faster operation, but to achieve this advantage there can be essentially no skew between the

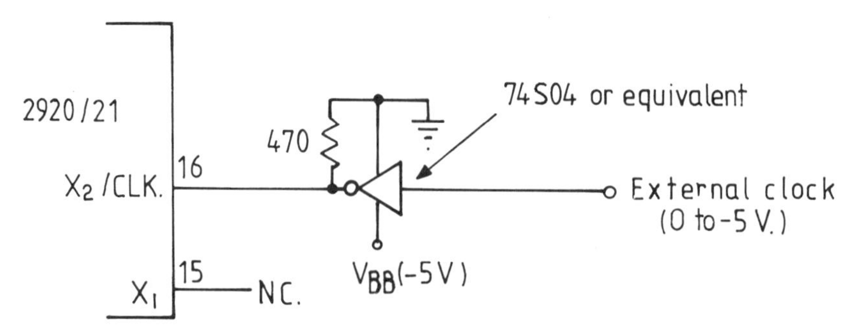

Figure 3.11 Clock

edges of the two clock signals. This can be difficult to attain with off-the-shelf components.

References

3.1 *2920 Analog Signal Processor Design Handbook* (1980) Intel Corporation.
3.2 *Using the 2920 Signal Processor in Modem Applications* (1981) Intel Corporation.

CHAPTER 4
THE NEC 7720

A. Zoicas, NEC Europe

4.1 Introduction

The increasing interest in digital signal processing (DSP) is due to the availability of powerful and cost-effective VLSI devices able to implement efficient DSP algorithms. The NEC μPD 7720 Signal Processing Interface (SPI) (1) is a microprogrammable device, and is the subject of the present chapter. Emphasis will be laid upon its pipeline structure, parallel processing and easy system implementation. These will be reflected by the implementation of classical examples of IIR/FIR filters and FFT programs.

As speed and algorithm implementation efficiency is to some degree limited for any given device, it appears reasonable to create a family of devices. There are two major directions to follow. The first is to keep the general-purpose character by modifying speed, wordlength and calculation accuracy, memory size and management, parallelism and pipelines, input–output flexibility and synchronisation, etc. (refs. 4.1, 4.2, 4.3). The second direction is to tailor the processors' architectures for specific algorithms like FFT, LPC, Dynamic Programming, etc. (refs. 4.4, 4.5). The family concept for Digital Signal Processors will be complete only if their surroundings are also considered. Easily interfaceable A/D and D/A converters, multiplexers/demultiplexers or compact analog interfaces (refs. 4.4, 4.6, 4.7) round out the family picture.

4.2 7720 Description

The design concept was to define a flexible and versatile microprogrammable number cruncher as an intelligent peripheral able to work also as a stand-alone processor. Although aimed mainly for telecommunications, where there are stringent requirements on accuracy, speed and signal interface compatibility with existing systems, the 7720 SPI has a more general character which makes it suitable for a broader range of applications. A parallel arrangement that provides concurrently a number of arithmetic and memory access operations and fetching and decoding of the next instruction, are the highlights of the 7720's Harvard architecture (fig. 4.1). Both 7720 mask- and 77P20 electrically-programmable SPIs, were fabricated with the most advanced 3 μm n-channel E/D MOS technology.

4.2.1 7720 Architecture

The processor is controlled by directly executable 23-bit horizontal micro-instructions. The fetching and decoding of the next instruction is done concurrently with the execution of the present instruction. One external 8 MHz clock is divided internally into more phases to control the internal pipeline and parallelism, giving a 250 ns instruction cycle time.

Memories are considered below in four sections:

Instruction-ROM of 512 $\times$ 23-bits reflects by its wordlength the simultaneous multi-operation capability. A four level stack allows the implementation of nested program loops.

Data-ROM of 512 $\times$ 13-bits may store fixed data: filter coefficients, look up tables, FFT windows and bit reversal data, etc. The ROM is indirectly addressable through one pointer (RP), which can be post-decremented as part of an instruction.

Data-RAM of 128 $\times$ 16-bits is broken into two halves; each half is addressed by the same pointer DP (fig. 4.2). The most significant bit (MSB) of the DP is not used for addressing, but decides which of the two halves interfaces to the internal data bus (IDB). In either case, on a read, the high RAM value is always made available for input to the K-register of the multiplier. This unique feature makes it possible for the same, or two different, RAM values to be sent to the multiplier in the same instruction. An additional sub-bus is available for direct access to the ALU input. The RAM is indirectly addressable and the DP can be post-modified during the execution of an instruction. As part of the same instruction effecting a load-multiply and add, two distinct parts of the DP can be modified in different ways. The high 3-bits (DPH) are modified by being exclusive-ORed with a three-bit pattern in the instruction word. The low 4-bits (DPL) can be

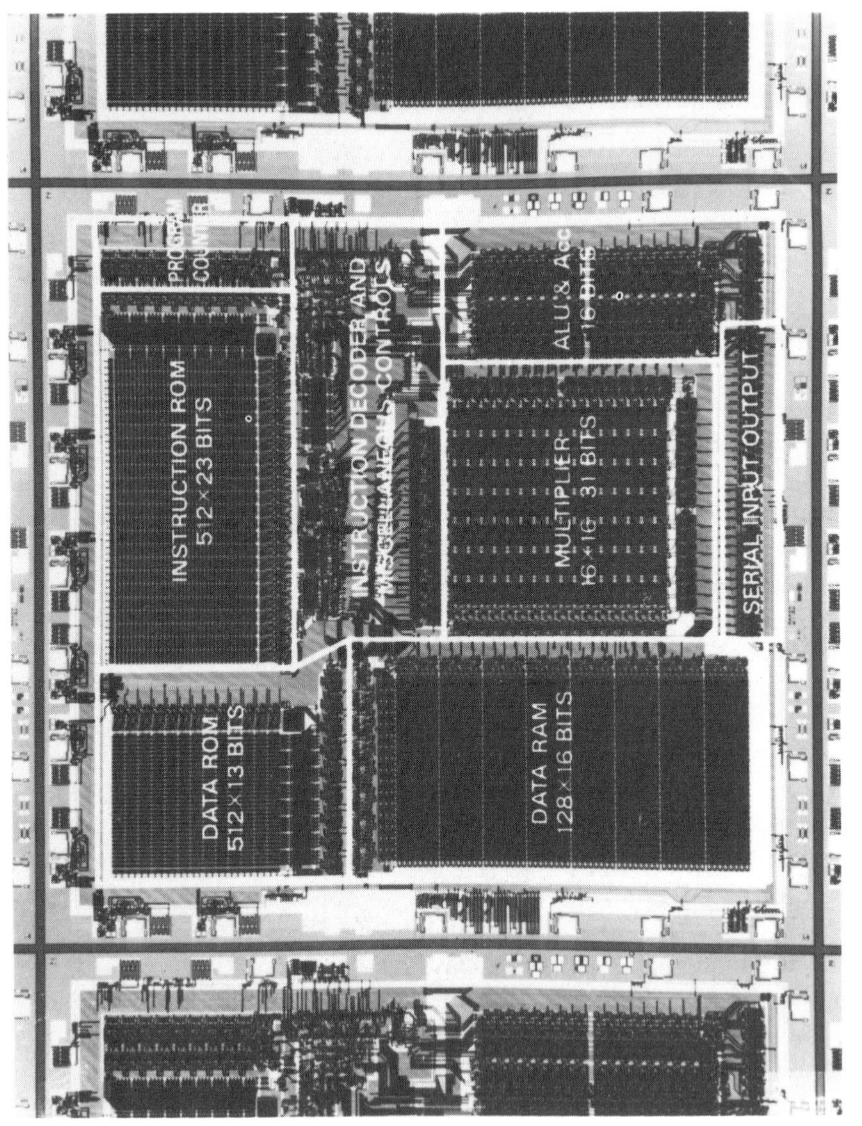

Plate 4.1 NEC 7720 bar photograph

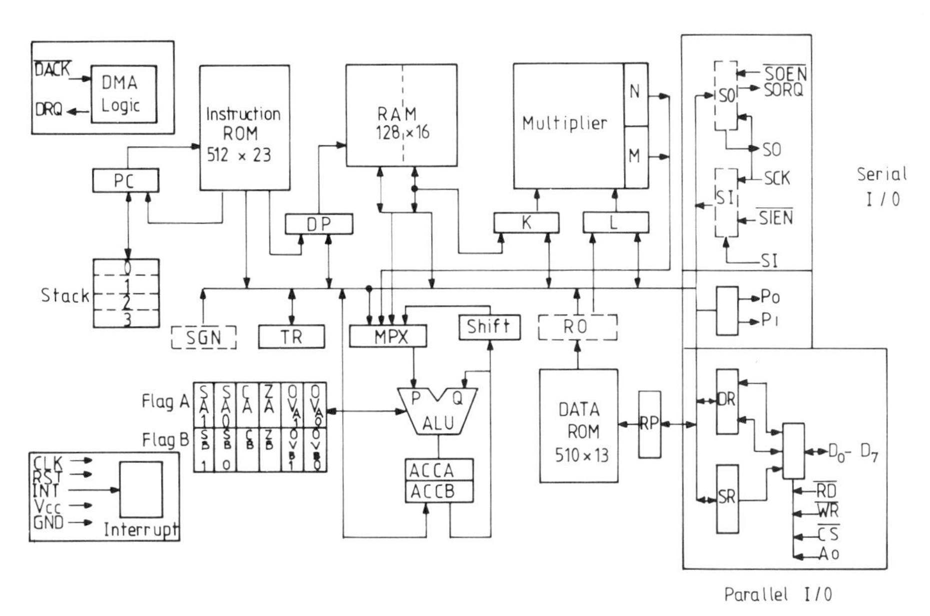

Figure 4.1 Block diagram

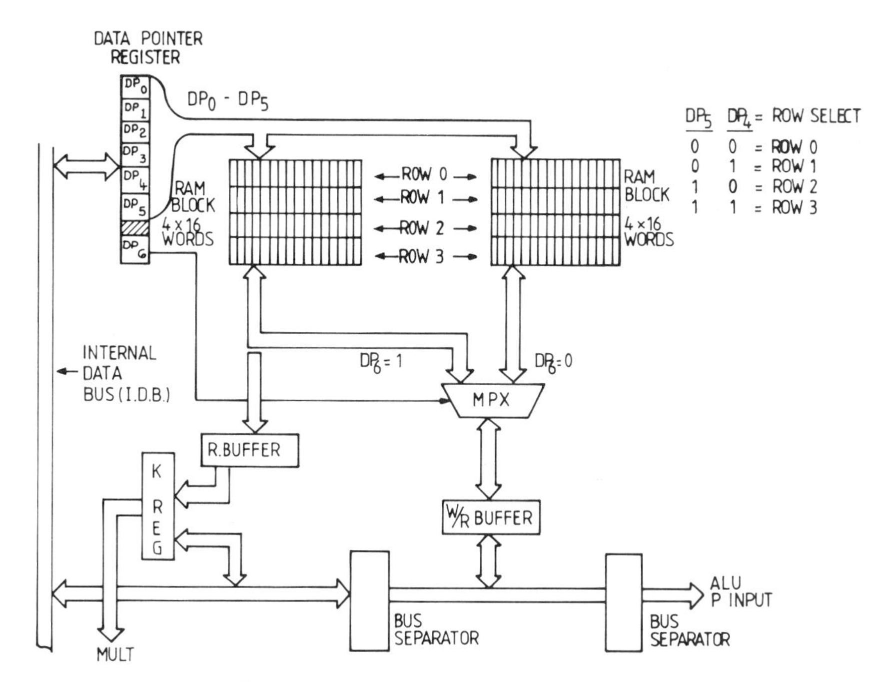

Figure 4.2 Data RAM and peripherals

incremented, decremented, cleared or not affected depending on two other bits in the instruction word. This partitioning of DP gives the RAM a column/row structure, where the row is defined by DPH and the column by DPL. An additional feature arises from this structure: by placing all the values for an individual calculation in the same column (or row), a subroutine for that calculation can be independent of column (row) position, and therefore, a whole array of values requiring the same calculation could use one or two basic routines. The software controllable flags DPL0, DPLF for detection of row beginning/end together with the post-increment/decrement of DPL may be used as loop counters.

Scratch-pad memory for 16-bit words is also provided by one transition register (TR), the input registers of the multiplier (K and L) and the data register (DR). A special mnemonic DRNF should be used for DR in order not to effect the DMA request or the handshaking flag RQM.

4.2.2 Arithmetic and Logic Unit (ALU)

A two's-complement fixed-point 16-bit ALU (fig. 4.3) is provided with two accumulators (Acc A and Acc B) which can also handle complex data or double-precision calculations on 32-bits. Direct inputs from the Data-RAM, multiplier outputs (M, N) and left/right shifter, for data scaling, offer multiple sources to the ALU and avoid overloading the IDB. The availability of the accumulator not involved in an ALU operation for data transfers increases the flexibility of this unit. A very sensitive point in fixed-point arithmetic generally and in two's-complement representation especially, is the overflow of the ALU. What solution should be adopted? Automatic replacement of the result which overflowed by saturation constants is normally used, but generates important distortions. For this reason in the 7720 SPI a 'look-back' software overflow control was selected. It very often happens that an overflow is cancelled by an underflow and vice versa, so that the end result in the accumulator is correct. Four flags for each accumulator are able to remember the overflow status for up to three consecutive ALU operations. The flags generate the correct saturation constant (+max or −max) in the sign register (SGN) after each operation. By testing the overflow status it is possible to replace the erroneous accumulator result by the SGN register content during the following instruction.

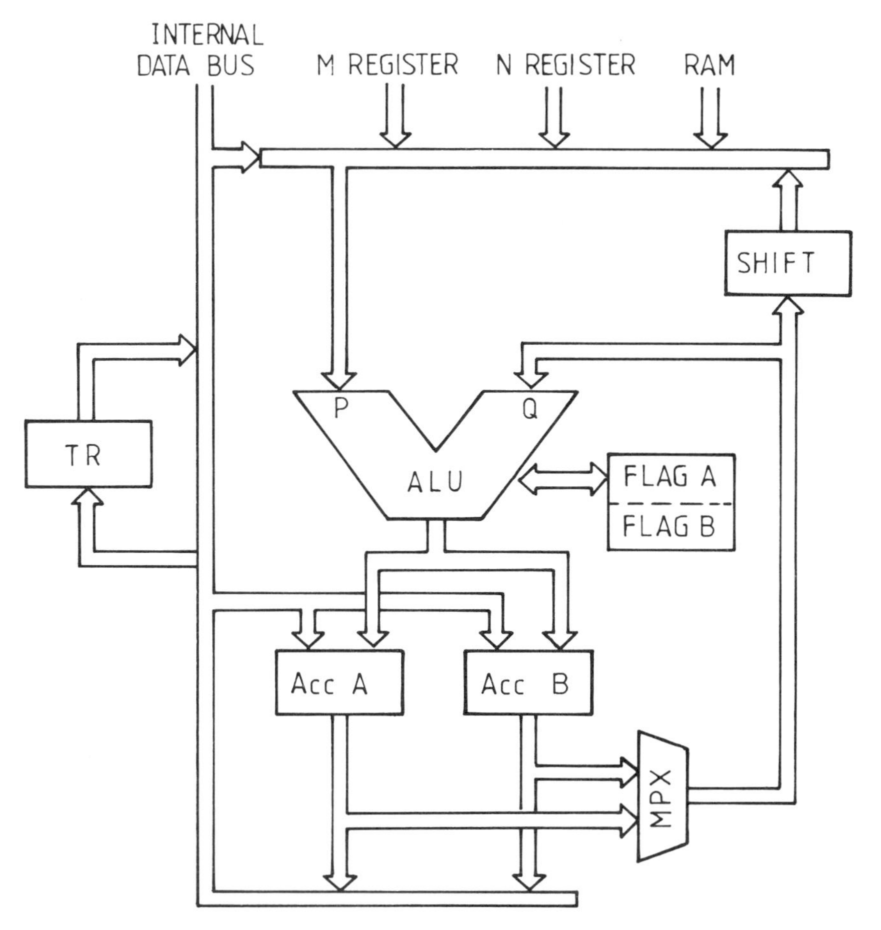

Figure 4.3 ALU and peripherals – block diagram

4.2.3 Multiplier

A hardware multiplier of the parallel–parallel type performs one $16 \times 16 \rightarrow 31$-bit signed-multiplication in one instruction cycle. This was possible by employing a carry-look-ahead technique using a modified Booth's algorithm. The input data range is assumed to be $[-1, +1)$. Hence, the hypothetical binary point is located after the first (sign) bit and this unit can not overflow. The multiple bus structure of the 7720 SPI – one main bus (IDB) and four sub-buses – allows simultaneous loading of the multiplier's input registers K and L from various locations. The full precision of the product in the M and N registers may be accumulated in two instruction steps by using another sub-bus.

4.2.4 Input/Output (I/O)

Various flexible and fast I/Os are an important feature of a digital signal processor, easing its system integration.

The 7720 I/O systems, conceived as slaves (figs. 4.4 to 4.6), are software-configurable to 8 or 16-bit wordlength. Flags are provided for internal checking of the I/O status, which allows an efficient synchronisation with the internal program flow.

Separate, double buffered, serial input and output channels provided with corresponding handshaking lines, cover at over 2 Mbps both CCITT and AT&T standard specifications for full-duplex PCM links. The possibility of reversing the order of the received data, during the transfer between the IDB and SO or SI registers, increases the number of serial oriented devices which can be connected to the 7720 (ADC, DAC, Codec, FIFO, etc.).

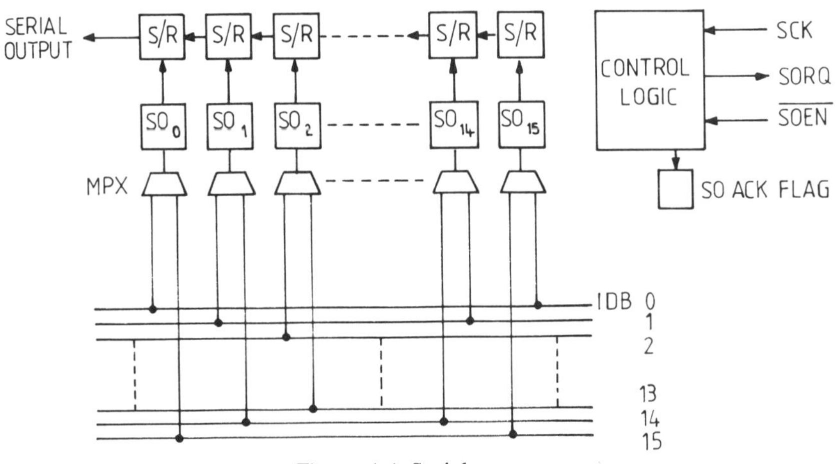

Figure 4.4 Serial output

The 8-bit parallel I/O port may be used for transferring data or reading the status of the SPI. Data transfer is handled through a 16-bit data register (DR) that is software-configurable to single or double byte transfers. Flags for handshaking and control lines for DMA define the two transfer modes. A special mnemonic DRNF allows exit from a transfer loop by inhibiting the setting of the RQM flag or a new DMA request.

A general purpose two-bit output port P0, P1 may be used for efficient signalling to external hardware.

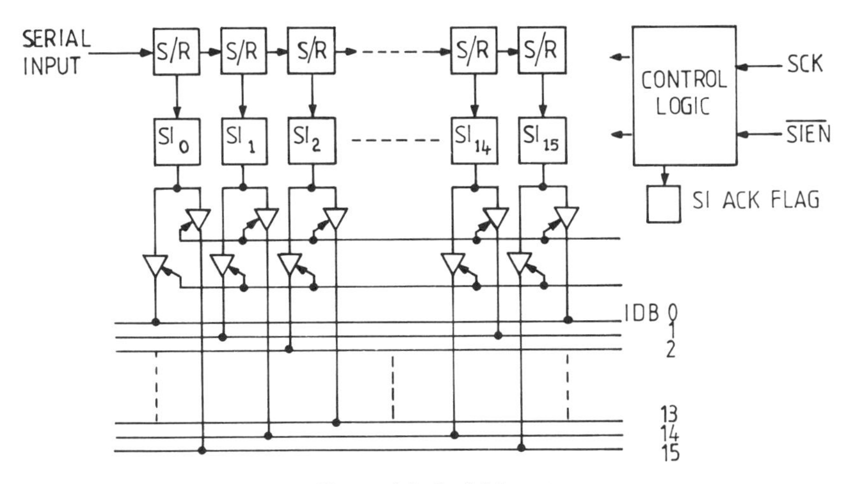

Figure 4.5 Serial input

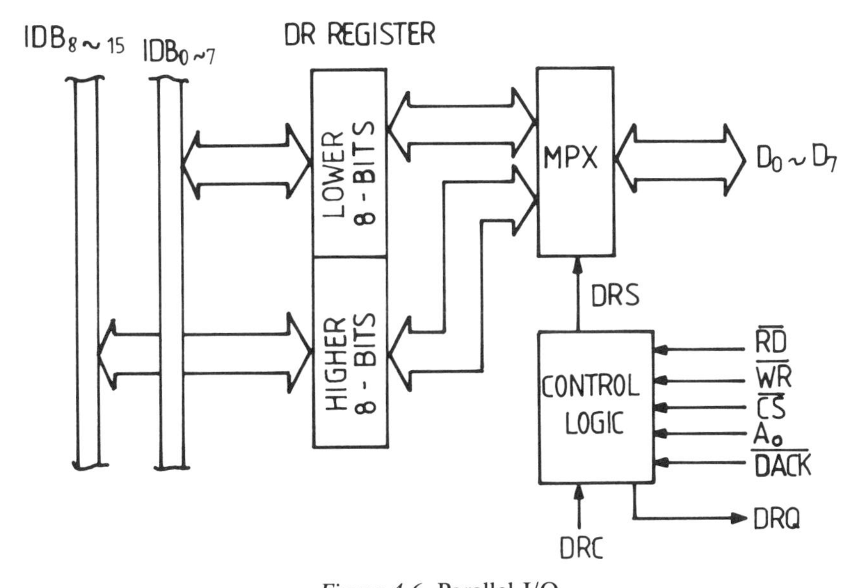

Figure 4.6 Parallel I/O

4.2.5 Interrupts

A one-level external interrupt uses a fixed vector pointing to the middle of the instruction-ROM. When used to synchronise the SPI with the sampling rate, an interrupt enable flag EI allows the possibility of ignoring a request as long as, for instance, a background routine is not completed.

4.3 7720 Instruction Set

All the instructions need only one cycle of 250 ns at an 8 MHz clock cycle. During the execution of the present instruction, the fetching and decoding of the next one is done in parallel. Input–output functions are carried out independently of the internal program flow. The 7720 SPI has three basic instruction types. The arithmetic-move (OR/RT) is the workhorse (program 4.1), which when used efficiently, allows up to seven tasks to be done in one instruction cycle, concurrent with I/O transfers.

The second instruction is the branch type and includes CALL, JUMP and 32 conditional jumps based on accumulator flags, I/O status or even the RAM pointer value (DPL).

The third and last type is load immediate data (LDI) to a destination using the IDB. This means that the instruction ROM can also be used to store fixed data and addresses.

```
IN PARALLEL:
                  - TRANSFER DATA VIA THE INTERNAL DATA BUS
                  - MULTIPLIER SOLUTION FROM CONTENTS OF MULTIPLIER
                    INPUT REGISTERS AVAILABLE
                  - EXECUTE ALU FUNCTION
                  - DATA RAM AND ROM POINTER MODIFICATION
                  - SUBROUTINE RETURN
EXAMPLE:
            OP MOV   @KLM,SI/ *  RAM (DP+64)→K, SI→L                 */
               ADD   ACCA,M / *  M+ACCA→ACCA                         */
               DPINC        / *  INCREMENT DATA RAM POINTER (DP)     */
               M1           / *  XOR 001 WITH HIGH 3 BITS OF DP      */
               RPDEC        / *  DECREMENT DATA ROM POINTER (RP)     */
               RET   ;      / *  SUBROUTINE (OR INTERRUPT) RETURN    */
```

All 7 functions are executed in one instruction cycle.
All DP and RP manipulation is done at the end of the instruction cycle.
One multiplication is parallely done during each cycle.

Program 4.1 Arithmetic-move instruction

4.4 7720 Development Process

4.4.1 System Integration of 7720 SPI

The layout of the 7720 reveals its peripheral character (fig. 4.7). Stand-alone operation is also possible (fig. 4.8). If high sampling rate or the computational task exceeds the capabilities of one 7720, it is very easy to cascade them (fig. 4.9). The peripheral character is its

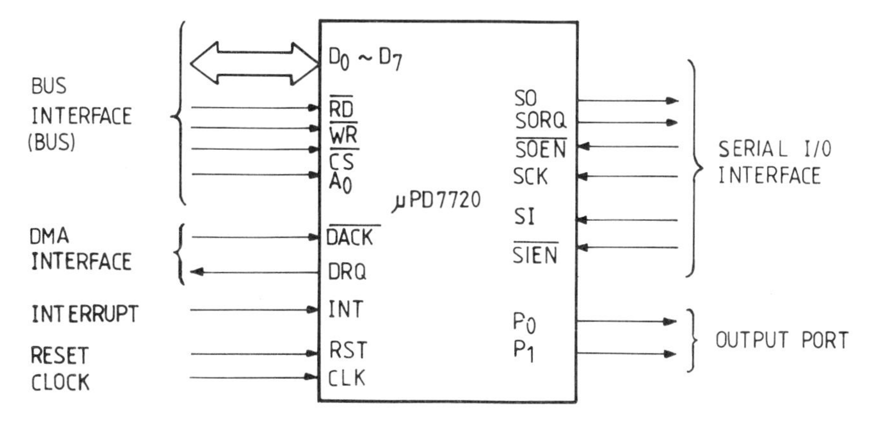

Figure 4.7 Layout

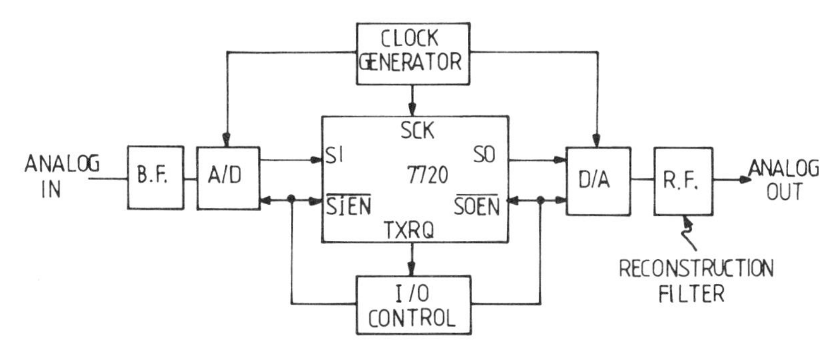

Figure 4.8 Stand-alone single processor

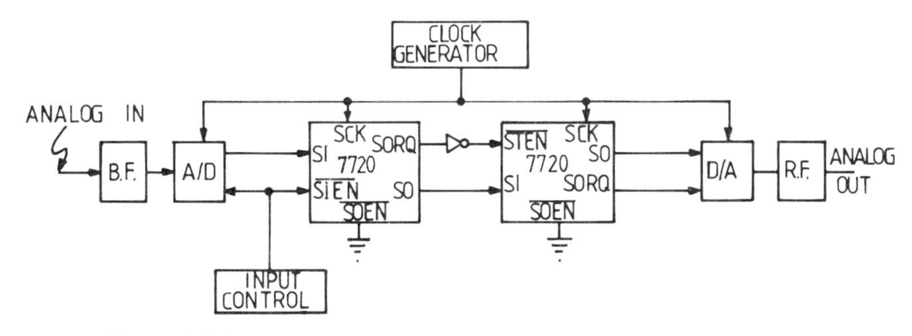

Figure 4.9 Two processors in cascade, increased input signal rate

major advantage when multiprocessor configurations of the serial-parallel type (fig. 4.10), or star connection (fig. 4.11) etc., have to work under the control of a CPU, sharing a common memory and performing either multiprocessing or parallel algorithms.

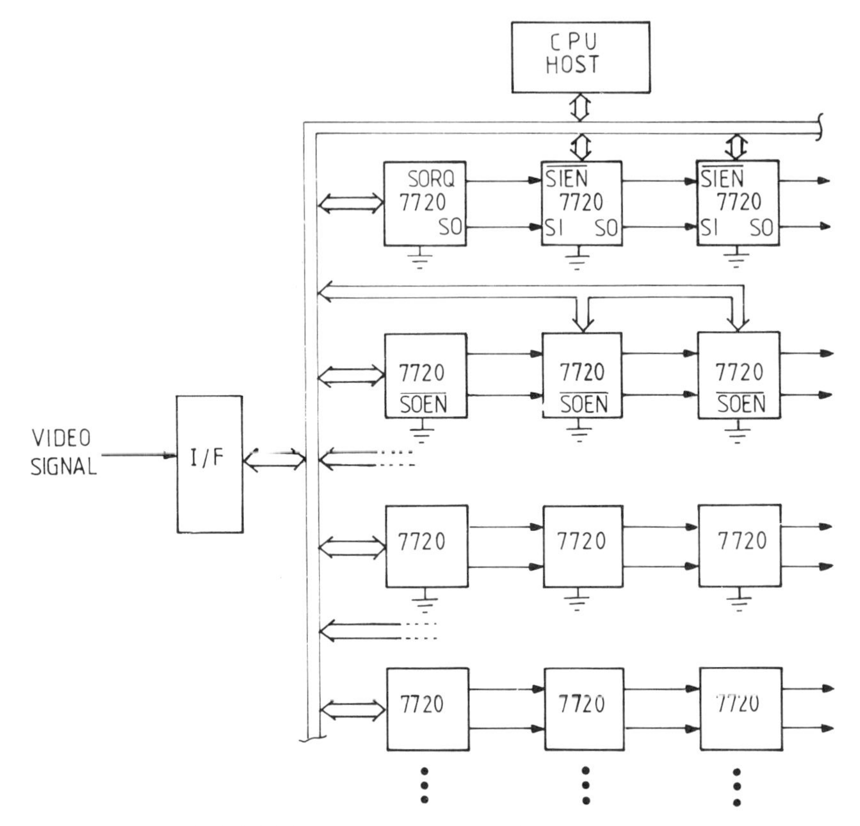

Figure 4.10 Multiprocessor structure of the serial–parallel type

This character is mandatory in high speed modems, radar and sonar applications, where 'good team work' without too much arbitration from the CPU side is indispensable.

4.4.2 Algorithm Implementation on 7720 SPI

Initially, the digital signal processing algorithm, written in a high level language, should be tested on a mainframe host computer. Target results are generated. The designer then translates the error-free algorithm into 7720 assembly language (fig. 4.12).

This source code is assembled by two programs (instruction- and Data-ROM), which generate the directly executable microprogram. Various hosts can be used: Intellec® II, PDP-11 or VAX families and finally microcomputer development systems supporting the CP/M® operating system.

Debugging of the 7720 microprogram can be done with a software

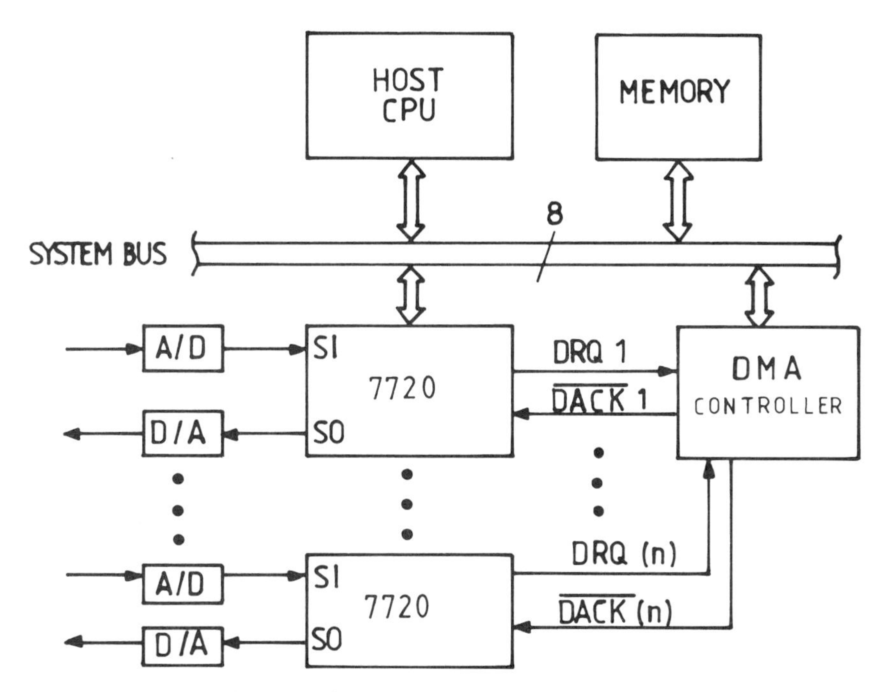

Figure 4.11 System using 7720s as complex computer peripherals

simulator to detect algorithm to microprogram translation errors by comparing the actual results generated with the target ones. Discrepancies between target and actual results require a new algorithm translation into 7720 assembly language.

In the following step the correct microprogram is tested in real-time by using the hardware emulator EVAKIT-7720. At this stage the hardware errors may be detected.

Finally, the bug-free microcode is burned into 77P20 EPROM SPIs and we can proceed to subsequent investigations on the target system prototype.

For low volume production of the final equipment or when different programs are executed by the SPIs, the 77P20 EPROM version should be selected. For medium volume production (a few hundred) the 7720 mask version will be the most economical solution.

Before deciding which digital signal processor should be used in a project, a careful evaluation of the devices available on the market is advisable. Own benchmarks or relevant test programs will simplify the decision making.

Powerful and convenient evaluation tools are available; such as the ISPS package from the Carnegie-Mellon University (refs. 4.8, 4.9).

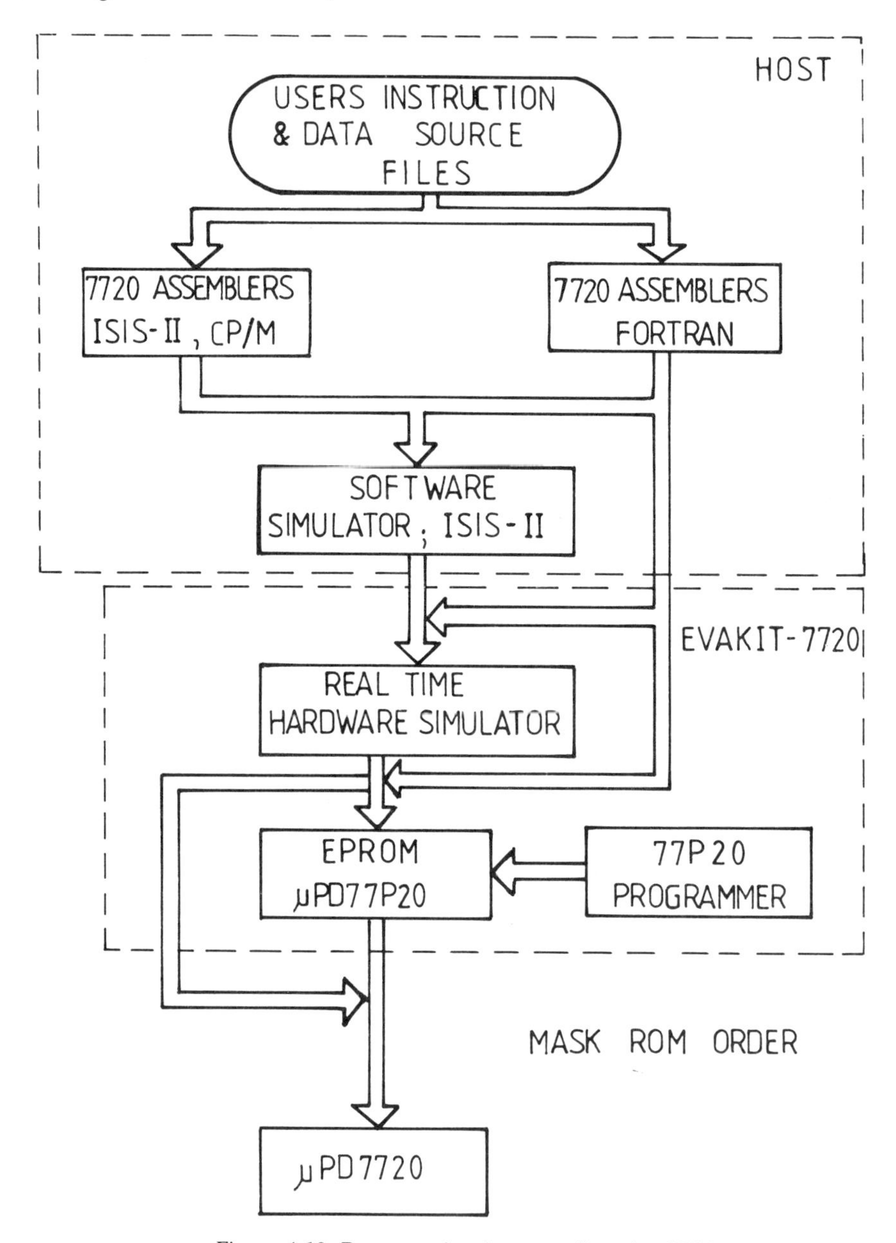

Figure 4.12 Program development flow for 7720

An excellent example of an application of this package, for the evaluation of the μPD 7720 of NEC, the TMS 320 of TI and the μPTS of CNET, is described in ref. 4.10.

The need to perform many operations in parallel and to provide a large number of parallel data paths creates major programming

Plate 4.2 EVAKIT – 7720

difficulties. The problems which have existed up to now seem to be solved when using high level language compilers to generate microcode for DSPs (ref. 4.11). A higher productivity during algorithm implementation can be achieved with such tools. The balance between installation cost and software development cost inclines towards the latter.

A comfortable, user-friendly and modular development system for single-chip digital signal processors was proposed by the Twente University in The Netherlands (ref. 4.12). This is no longer a mere proposal.

4.5 Application to Filters

The most common building blocks of digital processing algorithms, the IIR/FIR filters and the FFT will be highlighted.

Specific applications including LPC and channel vocoders, ADPCM Codec, DTMF receiver, modems and numerical control are treated in refs. 4.13, 4.14, 4.15, 4.16, 4.23, 4.24 and 4.25.

4.5.1 IIR Filters

Two implementations of a second-order (biquad) IIR filter stage will be presented. For an easy understanding of the advantages of using a device like the 7720 SPI, the direct canonic form (fig. 4.13) was selected.

Equations:

$$y_i = w_i + \alpha_1 w_{i-1} + \alpha_2 w_{i-2}$$
$$w_i = x_i - \beta_1 w_{i-1} - \beta_2 w_{i-2}$$

By cascading such stages, higher filter orders may be achieved and by connecting them in parallel, filter banks are easily obtained.

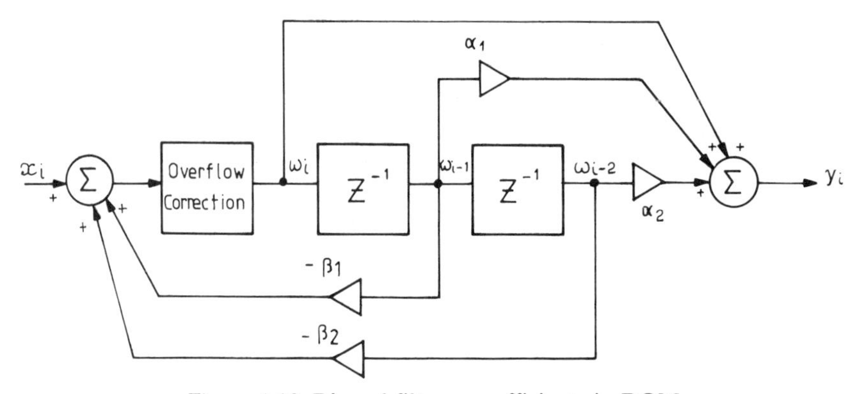

Figure 4.13 Biquad filter – coefficients in ROM

Independent scaling at the input and software overflow control after the calculation of the recursive section are common to the following routines.

(a) Biquad filter – coefficients in ROM (program 4.1)

The filter coefficients stored in Data-ROM were modified by adding or subtracting 1, in order to bring them into the processor's arithmetic range $[-1, +1)$.
Up to 64 stages (highest filter order 128) determined by the Data-RAM size may be cascaded and/or connected in parallel.
At 8 kHz sampling rate a total of 55 stages could be processed in real-time.

(b) Biquad filter – coefficients in RAM (table 4.1 and program 4.2)

The routine allows the filter characteristics to be changed at initialisation or during program execution, eventually adaptively. This time the coefficients larger than one have been halved in order to bring them into the 7720 arithmetic range (program 4.2).

$$Y(N) = W(N) + a(N1)W(N-1) + A(N2)W(N-2)$$
$$W(N) = X(N) - b(N1)W(N-1) - B(N2)W(N-2)$$
$$\text{where } B(N1) = b(N1)/2 \quad \text{and} \quad A(N1) = a(N1)/2$$

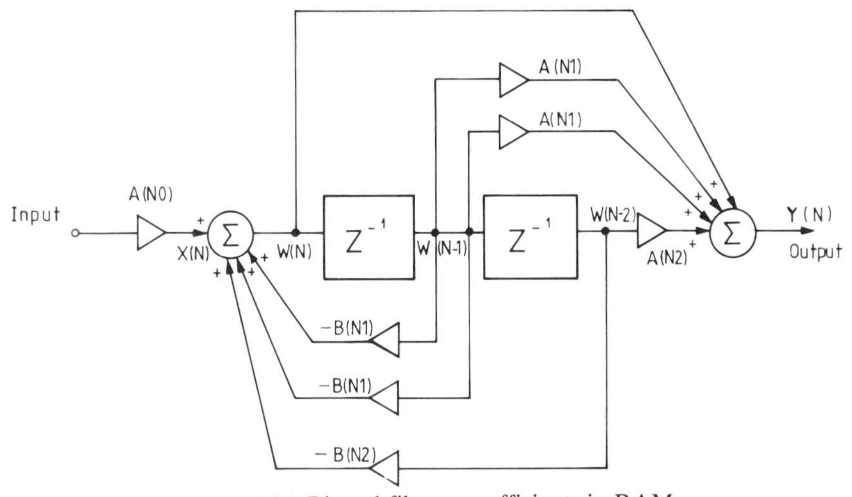

Figure 4.14 Biquad filter – coefficients in RAM

Table 4.1 Data RAM Map – biquad filter, coefficients in RAM

DP_H \ DP_L	0000	0001	0010	0011			1110	1111
000	W(F−1)	W(E−1)	W(D−1)	...	W(N−1)	...	W(1−1)	W(0−1)
001	W(F−2)	W(E−2)	W(D−2)	...	W(N−2)	...	W(1−2)	W(0−2)
010								
011	A(F0)	A(E0)	A(D0)	...	A(N0)	...	A(10)	A(00)
100	B(F1)	B(E1)	B(D1)	...	B(N1)	...	B(11)	B(01)
101	B(F2)	B(E2)	B(D2)	...	B(N2)	...	B(12)	B(02)
110	A(F1)	A(E1)	A(D1)	...	A(N1)	...	A(11)	A(01)
111	A(F2)	A(E2)	A(D2)	...	A(N2)	...	A(12)	A(02)

	DP_H			DP_L			
DP	6	5	4	3	2	1	0

$-2 < -\beta_1 < -1, \quad 1 < \alpha_1 < 2$

Initially: ACCA = x_i, OVA0 = 0, OVA1 = 0, RAM = w_{i-1}, ROM = $(-\beta_1 + 1)$

INST STEP	ALU OPERATION	REGISTER SET	NEXT ADDRESS	INSTRUCTION	
1	**ACCA ← ACCA − RAM** $/x_i/$ $/w_{i-1}/$	**K ← RAM, L ← ROM** $/w_{i-1}/$ $/-\beta_1+1/$	**RAM address modify** **ROM address decrement**	**OP**	**MOV @KLR,MEM** **SUB ACCA,IDB** **M1** **RPDEC ;**
2	**ACCA ← ACCA + M** $/x_i-w_{i-1}/$ $/(-\beta_1+1)w_{i-1}/$	**K ← RAM, L ← ROM** $/w_{i-2}/$ $/-\beta_2/$	**ROM address decrement**	**OP**	**MOV @KLR,MEM** **ADD ACCA,M** **RPDEC ;**
3	**ACCA ← ACCA + M** $/x_i-\beta_1 \cdot w_{i-1}/$ $/-\beta_2 \cdot w_{i-2}/$	**K ← RAM, L ← ROM** $/w_{i-2}/$ $/\alpha_2/$	**ROM address decrement**	**OP**	**MOV @KLR,MEM** **ADD ACCA,M** **RPDEC ;**
4	**JP TO 6 IF OVA1 = 0**			**JNOVA1**	**X ;**

	ALU			Data move	Address		Assembly	
5	ACCA ← SGN /+max or −max/					OP	MOV	@A,SGN ;
6	ACCA ← ACCA /w_i/	+	M /$\alpha_2 \cdot w_{i-2}$/	TR ← ACCA(OLD) /w_i/	RAM address modify	X: OP	MOV ADD M1 ;	@TR,A ACCA,M
7	ACCA ← ACCA /$\omega_i + \alpha_2 \cdot w_{i-2}$/	+	RAM /w_{i-1}/	K ← RAM, L ← ROM /w_{i-1}/ /$\alpha_1 - 1$/	ROM address decrement	OP	MOV ADD RPDEC ;	@KLR,MEM ACCA,IDB
8	ACCA ← ACCA /$w_i + w_{i-1} + \alpha_2 \cdot w_{i-2}$/	+	M /$(\alpha_1 - 1)w_{i-1}$/	RAM ← TR /w_i/	RAM address modify	OP	MOV ADD M1 ;	@MEM,TR ACCA,M
9				RAM ← K /w_{i-1}/		OP	MOV DPINC M1 RET ;	@MEM,K

Program 4.1 Biquad filter – coefficients in ROM

```
****************************************
*                                      *
*            FILTER ROUTINE            *
*                                      *
****************************************

OP      XOR     ACCA,IDB
        M3
        MOV     @K,A         ; /*CLR ACCA , INPUT TO K FOR SCALE, XOR DP-H & 011*
OP      M3
        MOV     @L,MEM       ;/*CHANGE DP-H TO 011*/
OP      ADD     ACCA,M
        MOV     @KLM,MEM     ; /*M+0=ACCA , W(N-1)*B(N1)*/
OP      ADD     ACCA,M
        M1
        MOV     @B,L         ; /*ADD FIRST TIME , CHANGE DP-H=001 , SAVE W(N-1)*/
OP      ADD     ACCA,M
        M6
        MOV     @KLM,MEM     ; /*ADD SECOND TIME B(N1)*W(N-1) , CHANGE DP-H=111 TO GET A (N2)
                                 B(N2)*W(N-2)*/
```

```
OP       ADD    ACCA,M
         MOV    @K,MEM        ;/*ADD TO ACCA , MOVE A(N2) TO K*/
JNOVA1   $+2                  ;/*OVERFLOW CHECK */
OP       MOV    @A,SGN        ;/*REPLACE BY SATURATION CONSTANT */
OP       ADD    ACCA,M
         M5
         MOV    @TR,A         ;/*ADD A(N2)*W(N-2) TO W(N) , CHANGE DP-H=010 , SAVE W(N)*/
OP       M2
         MOV    @KLM,B        ;/*W(N-1)=L , A(N1)=K , CHANGE DP-H=000*/
OP       ADD    ACCA,M
         M1
         MOV    @MEM,TR       ;/*W(N-1)*A(N1)=ACCA , CHANGE DP-H=001 , MOVE W(N) TO W(N-1)
OP       ADD    ACCA,M
         M1
         DPDEC
         MOV    @MEM,B
         RET                  ;/*ADD 2ND TIME, CHANGE DP-H=000, MOVE W(N-1) TO W(N-2) DECREMENT DP
                                 FOR NEXT STAGE & RESET*/
```

Program 4.2 Biquad filter – coefficients in RAM

Up to 16 stages can be implemented and the highest filter order of 32 is executed in real-time at sampling frequencies of over 20 kHz.

The design, realisation and evaluation of a 5th-order elliptic filter using a cascade of three 2nd-order stages should fix the methodology (ref. 4.17), (fig. 4.15 below and next two pages).

```
DIGITAL FILTER DESIGN
DIGF      :VERSION 2.4:
MARCH 1983

SAMPLING FREQUENCY Hz: 8000

1. A P P R O X I M A T I O N     - E L L I P T I C

                    | MENU |

          BIQU BUTT TSCH ITSCH ELL GU
          FIR READ STORE SCAL NEW GRAPH DISP PEM PRT
   SELECT: ELL
    ENTER: pass-band ripple db, stop-band att. db: .02,40
    ENTER: frequency: pass-band, stop-band Hz: 800,1280
    ENTER:  LP HP: LP
   SELECT: GO

    Z-DOMAIN BIQUADRATIC COEFFICIENTS: FULL PRECISION
```

SECTION	z^0 +	z^-1 +	z^-2	z^0 +	z^-1 +	z^-2
1	+1.00000	-0.21860	+1.00000	+1.00000	-1.18934	+0.49600
2	+1.00000	-0.97586	+1.00000	+1.00000	-1.35628	+0.83033
3	+0.00000	+1.00000	+1.00000	+0.00000	+1.00000	-0.55460

```
    ENTER: FREQUENCY RESPONSE ?: Y=with GR. D.  y=no GR. D.  S=SKIP: Y
    ENTER: FREQUENCY RANGE: min, step, max Hz: 50,50,4000
```

FREQUENCY Hz   TRANSFER FUNCTION GAIN IN DECIBELS

	0.0	50.0	100.0	150.0	200.0	250.0	300.0	350.0	400.0	450.0
50	+35.0	+35.0	+35.0	+35.0	+35.0	+35.0	+35.0	+35.0	+35.0	+35.0
550	+35.0	+35.0	+35.0	+35.0	+35.0	+35.0	+34.9	+34.1	+32.0	+28.3
1050	+23.8	+19.0	+13.9	+8.4	+2.0	-6.7	-43.5	-12.4	-9.0	-8.2
1550	-8.5	-9.5	-11.2	-13.5	-16.9	-22.3	-37.8	-26.7	-20.0	-16.5
2050	-14.3	-12.7	-11.5	-10.6	-10.0	-9.4	-9.0	-8.7	-8.5	-8.3
2550	-8.2	-8.2	-8.2	-8.2	-8.3	-8.4	-8.6	-8.8	-9.0	-9.3
3050	-9.6	-10.0	-10.4	-10.8	-11.3	-11.8	-12.3	-13.0	-13.7	-14.4
3550	-15.3	-16.3	-17.4	-18.7	-20.3	-22.2	-24.7	-28.2	-34.2	-144.6

```
    ENTER: m,M=MAGNITUDE  G=GROUP DELAY  P=PHASE  S=SKIP C=COPY: M
```

+35.02
-0.90
-36.81
-72.73
-108.64
-144.56

50.0   1050.0   2050.0   3050.0   4050.0   5050.0   6050.0

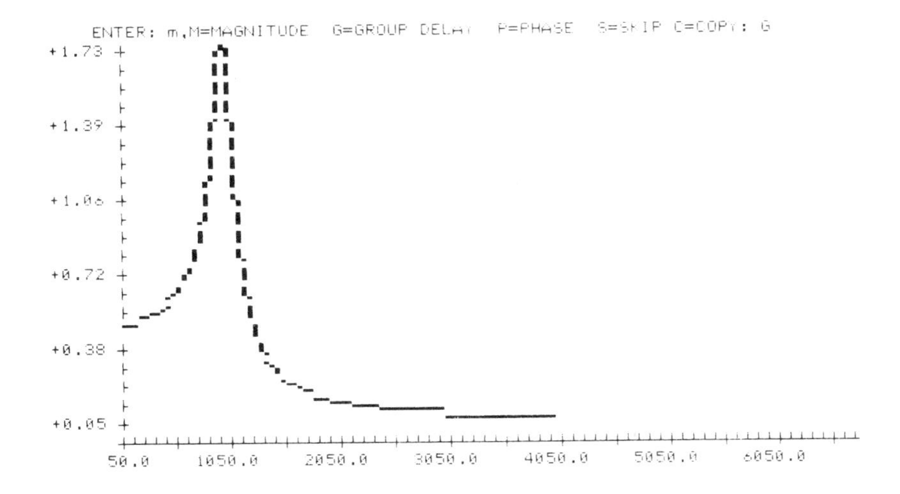

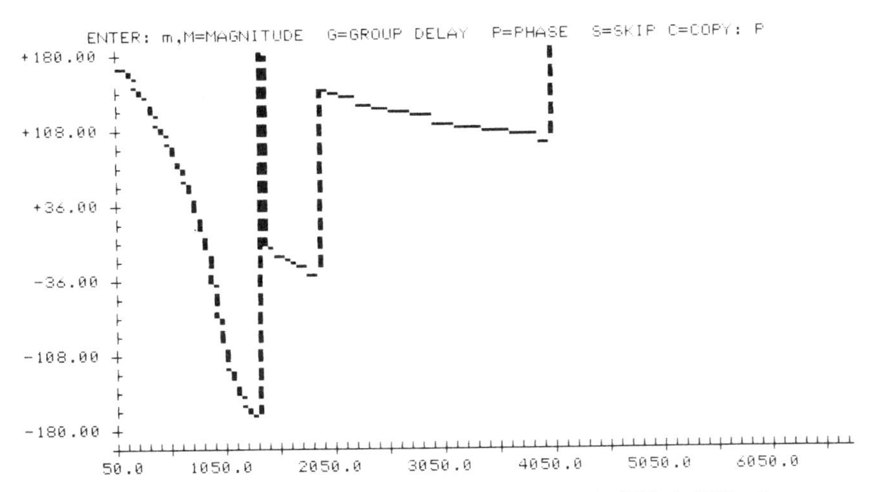

ENTER: m,M=MAGNITUDE G=GROUP DELAY P=PHASE S=SKIP C=COPY: S

2. R E A L I Z A T I O N
   THE DIRECT FORM WAS SELECTED (FIG.5.11)
   COEFFICIENTS AND SCALING FACTORS ARE 16 BIT LENGTH

3. S T U D Y O F A R I T H M E T I C E R R O R S
   FILTER VARIABLES (INPUT, OUTPUT, INTERMEDIATE) ARE 16 BIT LENGTH

4. I M P L E M E N T A T I O N

```
          UPD7720   RAM MAP WITH RECURSIVE COEFFICIENTS
SECTION    3   2   1
  A0      2000200020000000000000000000000000000000000000000000000000000000
  B1      237E56CD4C1E0000000000000000000000000000000000000000000000000000
  B2      000095B9C0840000000000000000000000000000000000000000000000000000
  A1      3FFFC18CF2030000000000000000000000000000000000000000000000000000
  A2      00007FFF7FFF0000000000000000000000000000000000000000000000000000
```

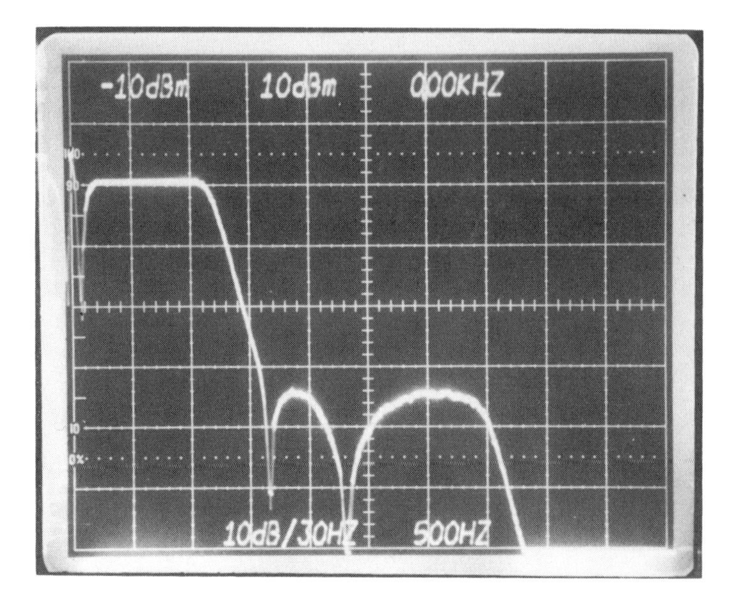

A 5th order elliptic filter

Satisfying:

Highest frequency in the passband	(f_a)	800Hz
Lowest frequency in the stopband	(f_p)	1280Hz
Ripple (p–p) in passband	(ε)	0.02dB
Minimum attenuation	(A)	40dB
Sampling frequency	(f_s)	8kHz

Figure 4.15

4.5.2 *FIR Filters*

Basic equation: $Y = \sum_{i=0}^{N-1} A_i \cdot X_i$

(a) Transversal FIR filter – coefficients in ROM.
By taking advantage of the impulse response symmetry, we save half of the required data-ROM locations for a given filter length N.

$$Y = \sum_{i=0}^{63} A_i(X_i + X_{127-i})$$

The length limitation of $N = 128$ taps comes from the Data-RAM size (fig. 4.16 and table 4.2). However, when the impulse response has a finer resolution than the delayed input samples,

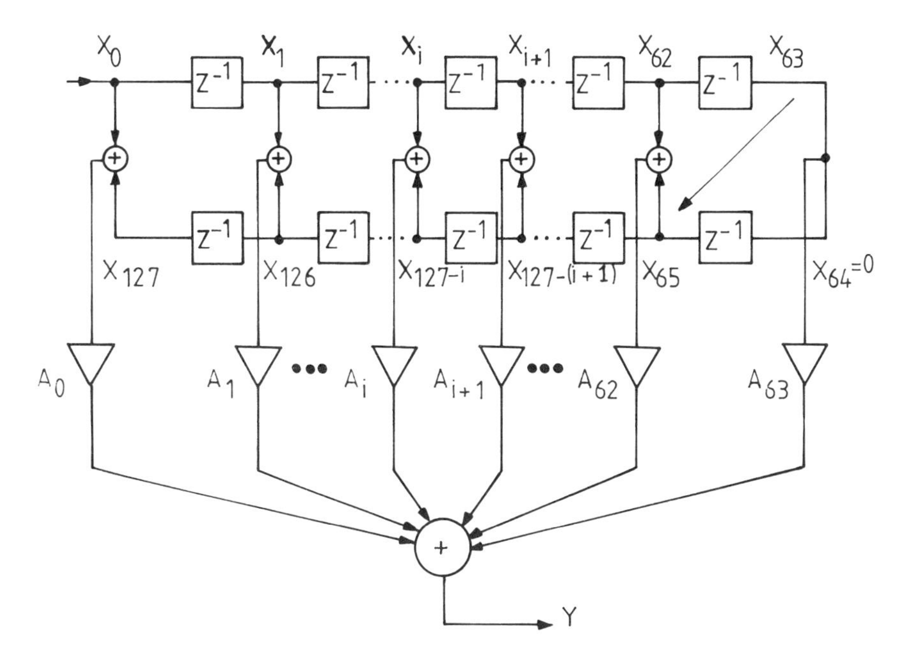

Figure 4.16 Transversal filter – coefficients in ROM

longer filters can be realised. A typical example of this is the shaping filter of a modem, where the delayed inputs are the symbols at the baud-rate and the impulse response is defined at the filter's sampling frequency.

One symmetric tap calculation including one sample delay and address preparation for the next tap takes only six instruction steps (program 4.3). Real-time execution including I/O functions are easily achieved at 8 KHz sampling.

Table 4.2 Data-RAM map – transversal filter, coefficients in ROM

DP_H \ DP_L	0000	0001	0010	...	...	1110	1111
100	X(0)	X(1)	X(2)	...	...	X(14)	X(15)
101	X(16)	X(17)	X(18)	.		X(30)	X(31)
110	X(32)	X(33)	X(34)	.		X(46)	X(47)
111	X(48)	X(49)	X(50)	.		X(62)	X(63)
000	X(127)	X(126)	X(125)	.		X(113)	X(112)
001	X(111)	X(110)	X(109)	.		X(97)	X(96)
010	X(95)	X(94)	X(93)	.		X(81)	X(80)
011	X(79)	X(78)	X(77)	.		X(65)	X(64)

```
                                  SUBROUTINE LOOP
                                  EACH INPUT SAMPLE WILL BE MULTIPLIED BY THE CORRESPONDING
                                  COEFFICIENTS A(I) AND A(128-I) AND THE RESULTS ADDED TO
                                  THE ACCA
LOOP:
                                        ACCA                ACCB            TR   DP(MEM)      MEM            K                L      RO

OP   MOV @B,MEM     ;                                       X(I)                 X(I)         X(I)                                   A(I)
OP   MOV @MEM,TR                                                                              X(I-1)
     M4             ;                                                            X(127-I)
OP   MOV @TR,B                                                              X(I)
     ADD ACCB,RAM                                       X(I)+X(127-I)
     DPINC          ;                                                            X(127-I+1)
OP   MOV @KLR,B     ;                                                                                     X(I)+X(127-I)   A(I)
OP   MOV @B,MEM                                         X(127-I+1)
     ADD ACCA,M                   A(I)*[X(I)+X(127-I)]
     DPDEC                                                                       X(127-I)
     RPDEC          ;                                                                                                            A(I-1)
OP   MOV @MEM,B                                                                               X(127-I+1)
     DPINC
     M4             ;                                                            X(I+1)

JDPL0 $+2           ;     IF DATA-RAM POINTER LOW =0 END OF LOOP
JMP  LOOP           ;
OP   M5                   MODIFY DATA-RAM POINTER
     RET            ;     RETURN OF THE MAIN PROGRAM
```

Program 4.3 Transversal filter – coefficients in ROM

(b) Transversal FIR filter – coefficients in RAM.
This routine uses the full precision of the multiplier's output and both accumulators to perform 32-bit multiply–add calculations. The filter output is rounded to 16 bits. Because of the RAM addressing, the saving of memory area by exploiting the symmetry of the impulse response comes at the expense of execution time. Hence, this routine does not exploit the symmetry.
The maximum filter length is $N = 64$ and for single precision taps only one instruction cycle is necessary. Using double precision taps takes three cycles (fig. 4.17, table 4.3 and program 4.4). Real-time execution including I/O functions for $N = 64$ can be achieved at over 12 kHz sampling.
The trade-off between program size (linear code) and execution time (branching) should be noted. The design, realisation and evaluation of a $N = 50$ FIR filter using the Remez exchange algorithm (ref. 4.18) is given in fig. 4.18.

Table 4.3 Data RAM map – transversal filter, coefficients in RAM

DP_H \ DP_L	0000	0001	0010	0011		1101	1110	1111
000		X(M-1)	X(M-2)	X(M-3)	..	X(M-13)	X(M-14)	X(M-15)
001	X(M-16)	X(M-17)	X(M-18)	X(M-19)	..	X(M-29)	X(M-30)	X(M-31)
010	X(M-32)	X(M-33)	X(M-34)	X(M-35)	..	X(M-45)	X(M-46)	X(M-47)
001	X(M-48)	X(M-49)	X(M-50)	X(M-51)	..	X(M-61)	X(M-62)	X(M-63)
100	A(0)	A(1)	A(2)	A(3)	..	A(13)	A(14)	A(15)
101	A(16)	A(17)	A(18)	A(19)	..	A(29)	A(30)	A(31)
110	A(32)	A(33)	A(34)	A(35)	..	A(45)	A(46)	A(47)
111	A(48)	A(49)	A(50)	A(51)	..	A(61)	A(62)	A(63)

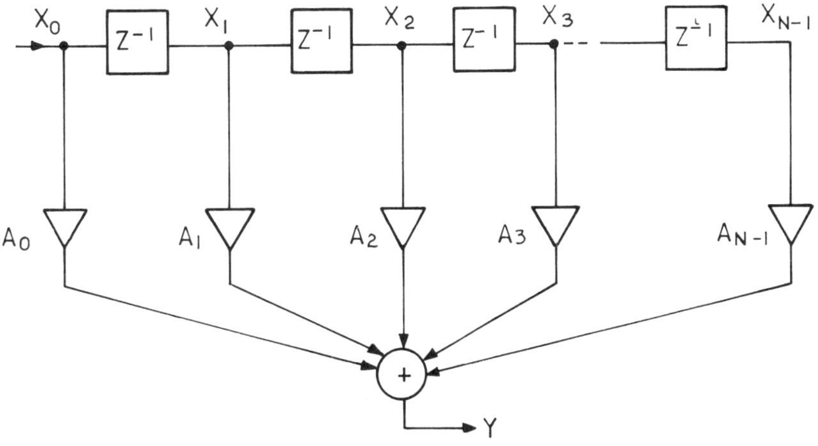

Figure 4.17 Transversal filter – coefficients in RAM

```
/* ****************************************************
   *                                                  *
   *          TRANSVERSAL FILTER SUBROUTINE           *
   *                                                  *
   **************************************************** */

TRFIL:
        OP      MOV     @KLM,MEM    ; /*MULT:DATA[X(M−I)]*COEFF.A[(I)] + ACCA = ACCA*/
        OP      ADD     ACCB,N
                MOV     @MEM,TR     ; /*SAVE X(M) RESP. X(M−I+1) TO MEM*/
        OP      ADC     ACCA,M
                MOV     @TR,L       ; /*SAVE X(M−I)*/
                DPINC
        JDPL0   $+2                 ; /*END OF PAGE ?*/
        JMP     TRFIL               ; /*IF NO CONTINUE*/
        OP      M1
                RET                 ;

/* ONE DOUBLE PRECISION 32-BIT TAP REQUIRES 6 INSTRUCTION STEPS */
```

Program 4.4 Transversal filter – coefficient in RAM

50 TAPS LINEAR PHASE FIR-BAND PASS FILTER

SATISFYING:

FILTER LENGTH	N = 50		
NO. OF BANDS	3		
	BAND 1	BAND 2	BAND 3
LOWER BAND EDGE	0.0	0.2	0.35
UPPER BAND EDGE	0.15	0.3	0.5
DESIRED VALUE	0.0	1.0	0.0
WEIGHTING	10.0	1.0	100.0
DEVIATION	0.0037	0.037	0.00037
DEVIATION IN dB	−48.62	−26.62	−68.62
EXTREMAL FREQ.....			

TAP	COEFFICIENT	HEX
A01 = A50	0.05648409E−02	0033
A02 = A49	0.30816298E−02	0064
A03 = A48	−0.31745254E−02	FF97
A04 = A47	−0.61980034E−02	FF34
A05 = A46	0.74350685E−02	00F3
A06 = A45	0.38368961E−02	007D
A07 = A44	−0.11103735E−01	FEB5
A08 = A43	−0.10101927E−01	FEB5
A09 = A42	0.89949206E−02	0126
A10 = A41	0.28980188E−02	005E
A11 = A40	0.26632992E−02	0057
A12 = A39	0.12021958E−01	0189
A13 = A38	−0.20657142E−01	FD5B
A14 = A37	−0.27189007E−01	FC85
A15 = A36	0.32337130E−01	0423
A16 = A35	0.28305613E−01	039F
A17 = A34	−0.20922041E−01	FD52
A18 = A33	−0.18761153E−02	FFC2
A19 = A32	−0.22823357E−01	FD14
A20 = A31	−0.53926217E−01	F918
A21 = A30	0.90472593E−01	DB94
A22 = A29	0.12315772E 00	0FC3
A23 = A28	−0.15639221E 00	EBFB
A24 = A27	−0.17733448E 00	E94D
A25 = A26	0.19078165E 00	186B

Figure 4.18 Design of optimal FIR filter using the Remez exchange algorithm

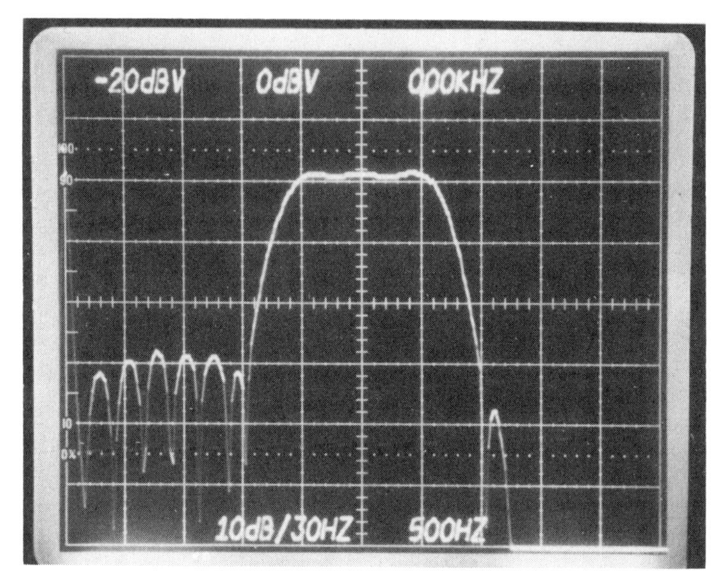

Figure 4.18 continued (Design of optimal FIR filter using the Remez exchange algorithm)

(c) Adaptive transversal FIR filter

Adaptive transversal filters have been used for channel equalisation in high speed modems for years. Without such a filter, high data rates over existing telephone lines would be nearly impossible. For this reason, telecommunications is one of the few industries using DSP techniques in their products, and has bitten the bullet of the cost of custom chips and expensive boards. The filter described below is a simplification of the one used in modems, functioning as a linear equaliser. Fig. 4.19 shows the algorithm structure. The equations of the three basic parts into which it can be broken for calculation purposes are given below:

1. Transversal filter equation

$$\hat{I}_{\kappa} = \sum_{i=-N}^{N} C_{i} V_{\kappa-i} \tag{1}$$

2. Coefficient adjustment

$$C_{(\kappa+1)i} = C_{\kappa i} + \Delta\varepsilon_{\kappa} V^{*}_{\kappa-i} \tag{2}$$

3. Detection and error

$$\tilde{I}_\kappa = \text{sgn}(\hat{I}_\kappa) \qquad i = -N, \ldots -1, 0, 1, \ldots, +N$$

$$\text{sgn}(x) = \begin{cases} +.7 + j.7, & R_e(x)>0, I_m(x)>0 \\ .7 - j.7, & R_e(x)>0, I_m(x)<0 \\ -.7 - j.7, & R_e(x)<0, I_m(x)>0 \\ -.7 - j.7, & R_e(x)<0, I_m(x)<0 \end{cases} \quad (3)$$

$$\varepsilon_\kappa = \tilde{I}_\kappa - \hat{I}_\kappa$$

Examining these equations, we will note that a multitude of multiply–add operations are necessary to implement this filter. The ability to do fast multiplies and adds is crucial to the use of any DSP component. This speed is directly related to the 'bandwidth' of the real-time application to which it can be applied. Another important aspect of this algorithm is that all the data and coefficients are variables and therefore must be kept in the data-RAM. An efficient technique accessing and addressing two variables for multiplication can be just as important to algorithm performance as the speed of multiplication and addition.

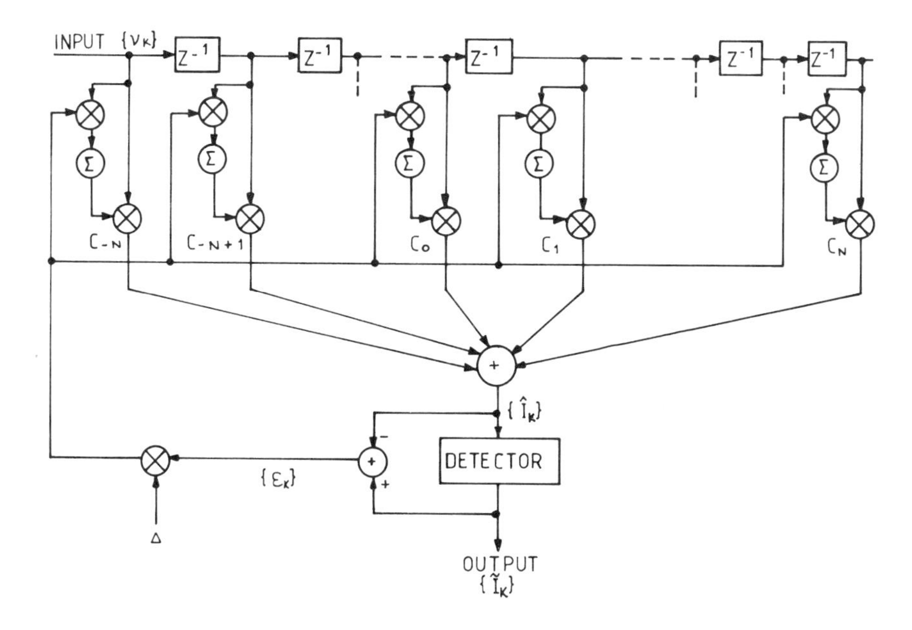

Figure 4.19 Adaptive transversal filter

The decision or detection part of the algorithm varies depending on the application. That is why it was left as a block in fig. 4.19. Most modem equalisers send the transversal filter's output to a controlling processor so that the processor can detect the symbol, make a limit value decision and send that decision back to the equaliser to calculate the filter error ε. A detection algorithm, using a hardlimited four state output (+/− 0.7, +/− j 0.7), completes our example (equation (3) above).

The individual operations of the tasks are: 4 address manipulations, 5 data accesses, 4 multiplies, 3 adds and 1 subtract to do the calculations necessary for just one of the taps in equation 1 using complex data and coefficients. In the example filter 24 taps of complex data and coefficients are used (table 4.4).

The calculation of a complex tap requires only four instruction cycles, and this gives some idea of the 7720 SPI performance. Program 4.5 shows the relevant section of code.

```
   /* ACCA = ACCB = 0,   REAL SUM IN ACCA, IMAG SUM IN ACCB
REAL DATA  = RD          IMAG DATA  = ID
REAL COEFF = RC          IMAG COEFF = IC
        INITIALLY DP = 0          */

OP      MOV     @KLM,MEM       /* K = RD, L = RC              */
        M5                   ; /* DPH = 101                   */
OP      MOV     @K,MEM         /* K = ID                      */
        ADD     ACCA,M         /* A = (RD*RC) + SUMR          */
        M4                   ; /* DPH = 0X1                   */
OP      MOV     @L,MEM         /* L = IC                      */
        ADD     ACCB,M         /* B = (ID*RC) + SUMI          */
        M5                   ; /* DPH = 1X0                   */
OP      MOV     @K,MEM         /* K = RD                      */
        SUB     ACCA,M         /* A = -(ID*IC) + SUMR         */
        M4                     /* DPH = 0X0                   */
        DPINC                ; /* DPL TO DPL+1, NEXT TAP      */
        /*GENERAL ROUTINES*/

OP      MOV     @KLM,MEM       /* K = RD, L = RC              */
        ADD     ACCB,M         /* B = (RD*IC) + SUMI          */
        M5                   ; /* DPH = 1X1                   */
/*AND SO ON .......*/
```

For complete information related to this algorithm, ref. 4.21 and 4.22 should be consulted.

Program 4.5 Section of adaptive transversal filter

DP_H \ DP_L	1	2	3	4	5	6	7	8	9	D	E	F
111	ID_{17}	ID_{18}	—	—	—	ID_{23}	ID_{24}					
110	RD_{17}	RD_{18}	—	—	—	RD_{23}	RD_{24}	$-\Delta$				
101	ID_1	ID_2	ID_3	—	—	—	—	—	------	ID_{14}	ID_{15}	ID_{16}
100	RD_1	RD_2	RD_3	—	—	—	—	—	------	RD_{14}	RD_{15}	RD_{16}
011	IC_{17}	IC_{18}	—	—	—	IC_{23}	IC_{24}	—				
010	RC_{17}	RC_{18}	—	—	—	RC_{23}	RC_{24}	—				
001	IC_1	IC_2	IC_3	—	—	—	—	—	------	IC_{14}	IC_{15}	IC_{16}
000	RC_1	RC_2	RC_3	—	—	—	—	—	------	RC_{14}	RC_{15}	RC_{16}

Table 4.4 Data RAM map – 24 tap adaptive transversal filter

4.6 Application to the Fast Fourier Transform (FFT)

Although not optimised for 'butterfly' calculations, the basic block of the FFT, the 7720 SPI performs it efficiently due to its powerful parallel processing. Calculation speed and accuracy, addressing efficiency and overflow dependent scaling in fixed-point arithmetic are major requirements for fast and correct butterfly processing.

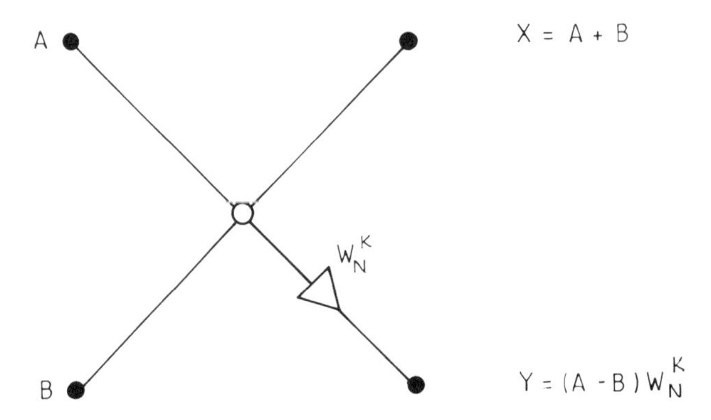

Figure 4.20 Butterfly for decimation in frequency

```
/*****     SUBROUTINE FOR BUTTERFLY COMPUTATION, REQUIRES          *****/
                        35 INSTRUCTION CYCLES

BFLY:  OP       XOR    ACCA,IDB    /* CLEAR ACCA */
                MOV    @NON,A      ;
J0:    OP       MOV    @DP,B       ;
       OP       ADD    ACCA,RAM    /* ACCA=XR */
                MOV    @K,MEM      ;/* STORE XR TO K-REG */
       OP       MOV    @DP,TR      ;
       OP       ADD    ACCA,RAM    ;/* ACCA=XR+YR */
       JNOVA0   J1                 ;/* JUDGE OVERFLOW */
       CALL     OVD                ;/* OVERFLOW CHECK */
J1:    OP       MOV    @DP,B       ;
       OP       XOR    ACCA,IDB    /* CLEAR ACCA */
                MOV    @MEM,A      ;/* STORE XR+YR TO RAM */
J2:    OP       ADD    ACCA,IDB    /* ACCA=XR */
                MOV    @NON,K      ;
       OP       MOV    @DP,TR      ;
       OP       SUB    ACCA,RAM    ;/* ACCA=XR-YR */
       JNOVA0   J3                 ;/* JUDGE OVERFLOW*/
       CALL     OVD                ;/* OVERFLOW CHECK */
J3:    OP       XOR    ACCA,IDB    /* CLEAR ACCA */
                MOV    @MEM,A      /* STORE XR-YR TO RAM */
                DPINC              ;
```

```
J4:     OP      ADD     ACCA,RAM     /* ACCA=YI */
                MOV     @DP,B        ;
        OP      DPINC                ;
        OP      ADD     ACCA,RAM     /* ACCA=YI+XI */
                MOV     @K,MEM       ;/* STORE XI TO K-REG */
        JNOVA0  J5                   ;/* JUDGE OVERFLOW */
        CALL    OVD                  ;/* OVERFLOW CHECK */
J5:     OP      XOR     ACCA,IDB     /* CLEAR ACCA */
                MOV     @MEM,A       ;/* STORE XI+YI TO RAM */
J6:     OP      MOV     @DP,TR       ;
        OP      ADD     ACCA,IDB     /* ACCA=XI */
                MOV     @NON,K
                DPINC                ;
        OP      SUB     ACCA,RAM     ;/* ACCA=XI-YI */
        JNOVA0  J7                   ;/* JUDGE OVERFLOW */
        CALL    OVD                  ;/* OVERFLOW CHECK */
J7:     OP      XOR     ACCA,IDB     /* CLEAR ACCA */
                MOV     @MEM,A       /* STORE XI-YI TO RAM */
                DPDEC                ;
J8:     OP      MOV     @KLR,MEM     /* (XR-YR)*WR */
                DPINC
                RPDEC                ;
        OP      ADD     ACCA,M       /* ACCA=(XR-YR)*WR */
                MOV     @KLR,MEM     /* (XI-YR)*WI */
                DPEC                 ;
        OP      ADD     ACCA,M       /* ACCA=(XR-YR)*WR+(XI-YI)* WI */
                MOV     @KLR,MEM     ;/* (XR-YR)*WI */
        JNOVA0  J9                   ;/* JUDGE OVERFLOW */
        CALL    OVD                  ;/* OVERFLOW CHECK */
J9:     OP      XOR     ACCA,IDB     /* CLEAR ACCA */
                MOV     @MEM,A       /* STORE (ACCA) TO RAM */
                DPINC
                RPDEC                ;
J10:    OP      SUB     ACCA,M       /* ACCA=-(XR-YR)*WI */
                MOV     @KLR,MEM     ;/* (XI-YI)*WR */
        OP      ADD     ACCA,M       ;/* ACCA=-(XR-YR)*(XI-YI)*WR */
        JNOVA0  J11                  ;/* JUDGE OVERFLOW */
        CALL    OVD                  ;/* OVERFLOW CHECK */
J11:    OP      INC     ACCB
                MOV     @MEM,A       ;
        OP      MOV     @A,TR        ;
        OP      INC     ACCA         ;
        OP      INC     ACCA         ;
        OP      INC     ACCB         /* NEXT ADDRESS OF RAM (ACCB) */
                MOV     @TR,A        /* NEXT ADDRESS OF RAM (TR) */
                RET                  ;
```

Program 4.6

The radix-2 butterfly decimation in frequency requires two complex additions and one complex multiplication (fig. 4.20, program 4.6)

$$W_N^K = \cos(2\pi K/N) - i \sin(2\pi K/N)$$

The generation of linear code, avoiding program branching shortens the execution time. An improvement of the execution time is expected also when using radix-4 or higher radices; this will shift the computational tasks towards address calculation. The 7720 SPI can

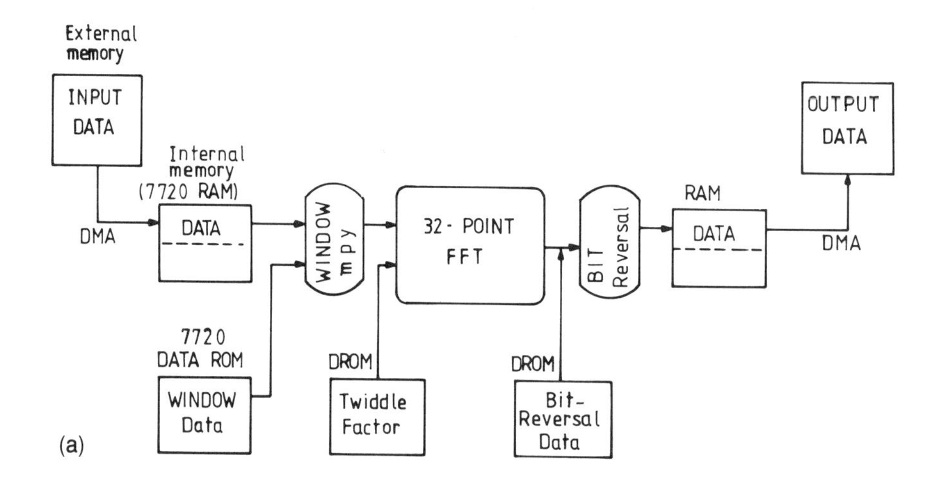

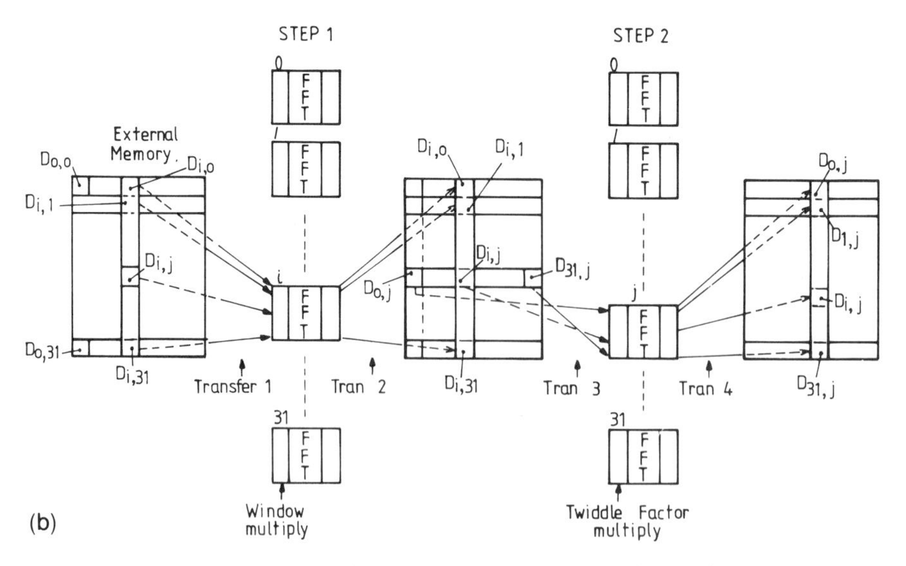

Figure 4.21 (a) 32-point FFT; (b) 1024-point FFT by repetitively using 32-point FFT

Table 4.5 Memory maps for 32 point and 1024 point FFT

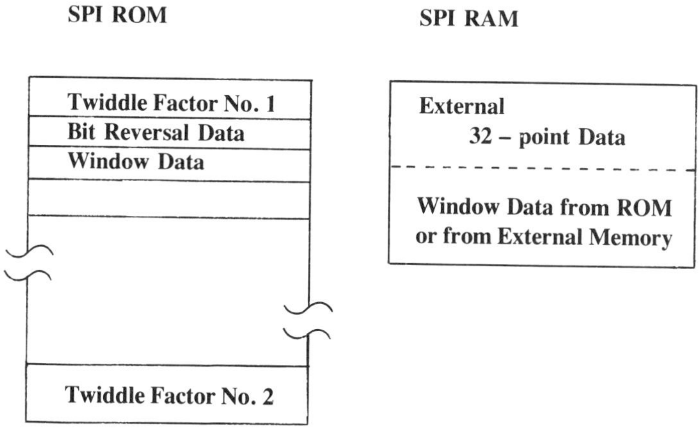

SPI ROM	SPI RAM
Twiddle Factor No. 1	External 32 – point Data
Bit Reversal Data	Window Data from ROM or from External Memory
Window Data	
Twiddle Factor No. 2	

process on chip up to 64 complex points FFT/IFFT, however, to realise higher orders the on-chip FFT size should be reduced to 32 points. (fig. 4.21, and table 4.5). Transforms involving multiples of 32 (64, 128, 256 . . .) up to 1024 complex points can be done on a multibus® board size (fig. 4.21).

A dynamic range of better than 60 dB and a real-time execution at over 13 kHz for 32 points and over 6.5 kHz for 1024 points (fig. 4.22, and table 4.6) are the board's benchmarks.

4.7 Conclusion

The NEC 7720 SPI together with the older Intel 2920, AMI S2811 and the younger Texas Instruments TMS 320 belong to the first generation of single-chip digital signal processors. The characteristics of this generation are clearly shown in fig. 4.22. NEC's device is already an industry standard in modems and speech processing applications.

Dedicated analog interfaces reduce the overhead for 7720 system integration and its peripheral character eases multiprocessor structures. The family concept is completed by dedicated DSPs (only for NEC internal use).

There are already clear ideas about a second generation of DSPs. Enhancement in data wordlength (eventually floating-point representation), on-chip memory capacity, speed, off-chip program-memory handling and user-defined MACROs are the highlights of NEC's future generation.

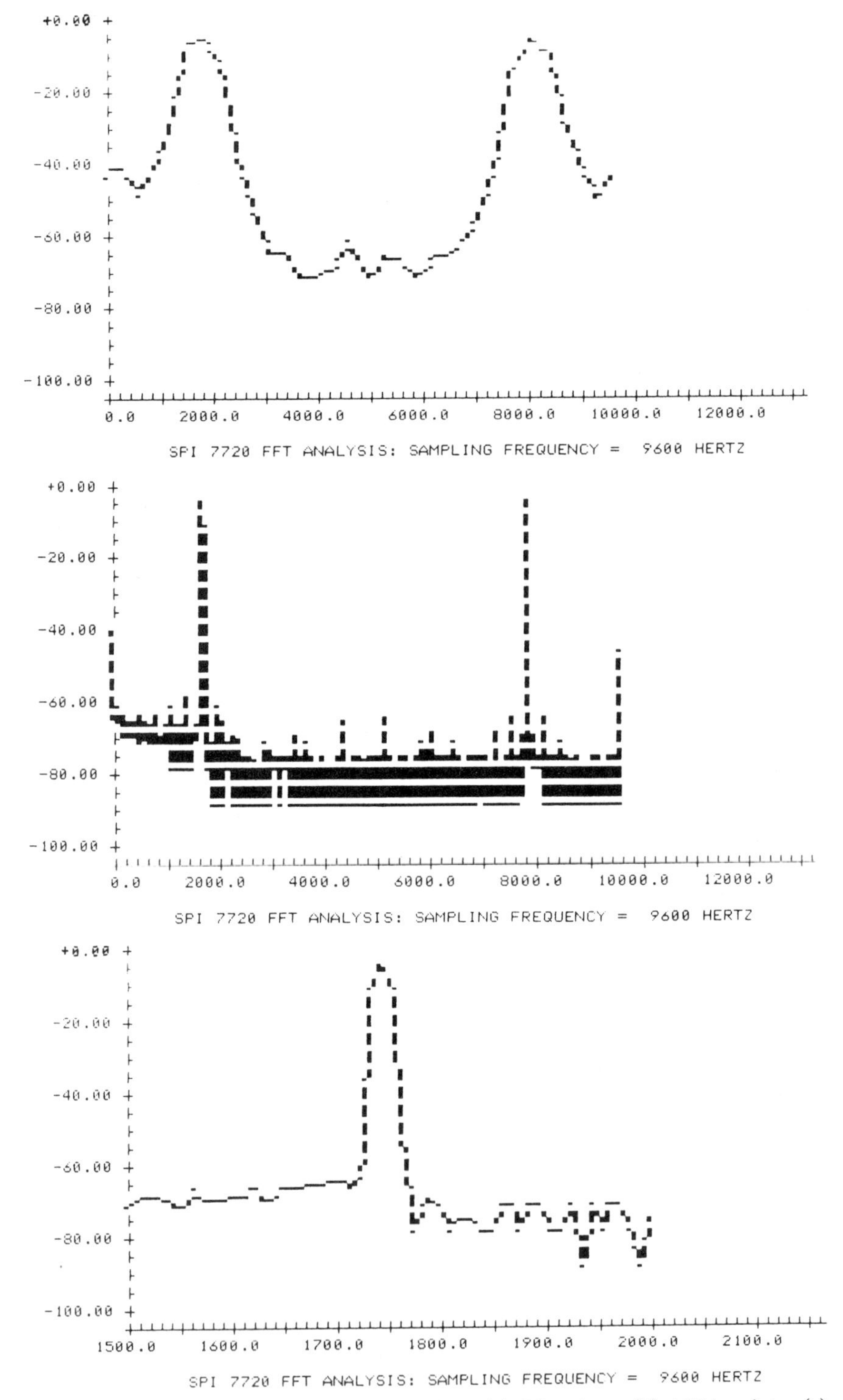

Figure 4.22 FFT analysis – sinusoidal function: (a) 32 points; (b) 1024 points; (c) 1024 points with frequency axis expansion

Table 4.6 FFT timing

32-POINT FFT	
Input	**163.8 μs (Memory → RAM)**
Window multiply	**48 μs**
Butterfly computation	**748 μs**
Bit Reversal	**89 μs**
Output	**163.8 μs (RAM → Memory)**
	1212.6 μs = 1.2 ms
WITH OVERFLOW PROCESSING AT EACH STEP	
BUTTERFLY COMP.	**1.16 ms**
TOTAL PROCESS TIME	**1.63 ms**
1024-POINT FFT	
Input	**10.5 ms**
Window	**1.5 ms**
Twiddle Factor	**3.4 ms**
TFM	**2.8 ms**
Butterfly	**47.9 ms**
Bit Reversal	**5.7 ms**
Output	**5.2 ms**
	77.0 ms
WITH OVERFLOW PROCESSING AT EACH STEP	
BUTTERFLY COMP.	**74,25 ms**
TOTAL PROCESS TIME	**103,4 ms**

References

4.1 Nishitani, T., Kawakami, Y. *et al.* (1980) 'LSI Signal Processor Development for Communications Equipment' Washington: IEEE Press, *ICASSP 80* **3** 386–389.

4.2 Yano, M. *et al.* (1982) 'An LSI Digital Signal Processor' Paris: IEEE Press, *ICASSP 82* **2** 1073–1077.

4.3 Watanabe, K. *et al.* (1978) 'A 4800 Bit/s Microprocesor Data Modem' Washington: IEEE Press, **5** 493–498.

4.4 Tanaka, H. *et al.* (1982) 'A Speech Recognition LSI Chip Set' *Munchen, Microelectronics Congress Proceedings*, 14–23.

4.5 Ishizuka, H. *et al.* (1983) 'A Microprocessor for Speech Recognition' New York: IEEE Press, *ISSCC 83* **2** 111–115.

4.6 Kuraishi, Y. *et al.* (1982) 'A Single-Chip NMOS Analog Front-end LSI For Modems' IEEE Press, *Journal of Solid-State Circuits* **6**, 1039–1044.

4.7 Iwata, A. *et al.* (1981) 'A Single-Chip CMOS PCM CODEC

With Switched-capacitor Filters' IEEE Press, *ISSCC 81 Digest of Technical Papers*, 244–245.

4.8 Barbacci, M. R. *et al*. (1980) *The Symbolic Manipulation of Computer Descriptions: An ISPS Simulator* Carnegie-Mellon University, Technical Report.

4.9 Barbacci, M. R. (1981) 'ISPS: The Notation and its Applications IEEE Transactions on Computers **C-30** 24–40.

4.10 Sorel, Y. and Wolf, P. (1983) 'Evaluation d'Architectures de Microprocesseurs de Traitment du Signal' *Nice: Neuvieme Colloque sur le Traitement du Signal et ses Applications*.

4.11 Jansen, I. H. (1980) 'PRODISP, a Flexible and Fast Digital Signal Processor, Programmable in a High Level Language' *EUSIPCO 1980* 363–368.

4.12 Herrmann, O. E. and Smit, I. (1983) 'A User-friendly Environment to Implement Algorithms on Single-chip Digital Signal Processors' *Erlangen*, *EUSIPCO '83* 851–854.

4.13 Feldman, J. A. and Hofstetter, A. M. (1982) *A Compact, Flexible LPC Vocoder Based on a Commercial Signal Processing Microcomputer* MIT Press, Technical Report 1982.

4.14 Feldman, J. A. (1982) 'A Compact Digital Channel Vocoder using Commercial Devices' IEEE Press, *Paris ICASSP 82 Proceedings* **3** 1960–1963.

4.15 Nishitani, T. *et al*. (1982) 'A 32 Kbit/s Toll Quality ADPCM Codec using a Single-chip Signal Processor' IEEE Press, *Paris ICASSP 82 Proceedings* **2** 960–964.

4.16 NEC Electronics Europe (1983) *Workshop-Digital Signal Processor μPD 7720* Duesseldorf.

4.17 NEC Electronics Europe (1982) *Digital Filter Analysis-DIGF with Signal Processing Interface Device μPD 7720* Duesseldorf.

4.18 McClellan, I. H. *et al*. (1973) 'A Computer Program for Designing Optimum FIR Linear Phase Digital Filters' *IEEE Transactions on Audio and Electroacoustics* 128–139.

4.19 Epstein, D. (1981) 'An Adaptive Transversal Filter using the μPD 7720' *Proceedings of U.S. National Electronics Conference*.

4.20 Dischinger, T. and Nielinger, H. (1983) *Untersuchungen am Signalprozessor μPD 7720 von NEC* Furtwangen, Diplomarbeit.

4.21 NEC Electronics Europe (1983) *Fast Fourier Transform Analysis with the Signal Processing Interface Device NEC μPD 7720* Duesseldorf.

4.22 Blasco, R. W. (1980) 'Evolution of the Single-chip Digital Signal Processing: Past, Present and Future' IEEE Press, *ICASSP 80* **1** 417–421.

4.23 Vary, P. (1983) 'On the Enhancement of Noisy Speech' *Erlangen*, *EUSIPCO '83* 327–330.
4.24 Roethe, E. (1983) 'Digital Single Sideband Modulation for Radiotelephony' *Erlangen*, *EUSIPCO '83* 523–526.
4.25 Le Tourneur, G., Maitre, X. and Petit, J. P. (1983) 'An Experimental Digital PCM Test Equipment' *Erlangen*, *EUSIPCO '83* 439–442.
4.26 Hedelin, P. and Kollberg, L. (1983) 'RELP Vocoding using NEC 7720' *Stuttgart*, *ECCTD '83* **1**.

CHAPTER 5
THE TMS 320

P. Strzelecki, Texas Instruments, U.K.

5.1 Introduction

TMS 320 is a generic term encompassing a family of high performance 16/32-bit single chip microprocessors, microcomputers and peripherals, along with hardware and software development and application support, that has been defined and marketed by Texas Instruments. The primary markets for this family are in the wide application base of real time digital signal processing (DSP).

The family made quite a dramatic entry into this functionally specific DSP area following the disclosure of the TMS 32010 and TMS 320M10, the first microprocessor and microcomputer members of the family, at ISSCC, San Francisco in February 1982. At ICASSP, Boston April 1983 the TMS 32010(M) featured in numerous papers and coincided with the full product announcement of devices and development support by Texas Instruments. Like most useful innovations the TMS 320 family arrived via an evolutionary route, the forebearers here being the ubiquitous general purpose single chip microprocessor and microcomputers. A key element of the TMS 320 strategy was to make it general purpose within the wide spectrum of DSP applications. The consequential larger number of applications as well as the individual high volumes will assist in reducing device prices to enable further pervasion for instance into the consumer market.

It was deemed important that radical changes to architecture and instruction sets were not necessary on the TMS 32010(M). Unlike other competitive devices a familiar style, 'user friendly', instruction set has been implemented. Horizontal instruction sets requiring large instruction words (in the extreme a bit per internal CPU control signal) yield high performance. However a more vertical (in the extreme a code per machine operation) instruction set has been implemented. Analysis shows that there are only a few key parallel operations that significantly impact system throughput.

The numerically intensive nature of DSP algorithms is reflected in the functional blocks of the TMS 32010(M). On a typical general purpose microcomputer approximately 10% of a chip area could be termed arithmetic. On the TMS 32010(M) approximately 35% is given to such functions. The inclusion of two shifters, a single cycle 16×16 hardware multiplier and 32-bit adder/accumulator account for most of this.

The two devices have some unique features which make them second generation products in this area. They can act as masters or slaves. That is they can provide total system processing functions, both housekeeping and DSP, as well as acting in a coprocessing mode to some systems host. The ability to address external memory without speed penalty has proved a winning feature. This gives two benefits. Full system prototyping and evaluation using off chip memories can be achieved before committing to an unalterable mask ROM part. The second advantage is that low volume applications or systems requiring frequently changed code can be viably tackled by these devices using the TMS 32010 microprocessor version.

It is quite practical to implement a system comprising off-chip RAM for program storage. This would allow the program to be changed during run time, e.g. changing constants representing filter tap weights following some change in operating conditions. The flexibility of using on-chip, as well as off-chip memory gives Texas Instruments the opportunity to offer standard ROM customisations of the TMS 320M10. A device containing the essential routines for a series of filters could be a pre-programmed on-chip. The application code off-chip would use this by, for example, supplying coefficients of the taps.

Another unique feature of these devices is the support for interrupt driven systems. Three pins are offered to give a vectored reset, a maskable hardware vectored interrupt and a software polled line. The first two pins have the same function as in any general purpose device. The software polled line is supported via the instruction set and is instrumental in allowing multiprocessing TMS 32010(M) systems to be implemented. The ability to increase system throughput by using more than one TMS 32010 again widens its application base, particularly so in certain image processing and spectral analysis projects perhaps currently using bipolar bit slice technology.

One of the greatest strengths of the TMS 320 family is the development support, software, hardware and literature.

XDS/320 (extended development support) comprises both host independent hardware and software that will run on a variety of host computers (e.g. VAX, IBM and TI). During an initial system assessment the evaluation board (EVM) requiring a terminal and power supply can be used. A compatible analog interface board (AIB) will allow A/D conversion of analog signals for processing by the EVM. Following processing the AIB provides D/A conversion to move back into the analog domain.

More sophisticated developments can be achieved using the XDS 320 Macro assembler, linker, simulator and emulator. The Macro assembler and linker allows assembler modules to be developed, converted to object, and linked for execution. Execution can take place either in the software environment of the simulator for quick debug or in hardware using the emulator, providing in circuit emulation of a TMS 32010(M) target system.

Design utility packages are also available, for example, for filter design, providing a user interface via corner frequency, ripple, roll off, etc. and producing 'best match' 320 object code as output.

The devices are gaining third party software support with products available for CP/M, PDP11, and HP64000 amongst others.

The TMS 32010 was used in the voice recognition, store and forward function offered by Texas Instruments on its professional computer announced during mid-1983. A lot of work has been done in the speech area but also in the telecom sector, particularly for modems. The general purpose nature of these devices make them also applicable to applications requiring a high speed general purpose processor, e.g. robotic arm trajectory calculation and control.

The TMS 320 family is developing with two peripheral devices providing programmable analog input and analog output. A certain degree of redundancy has been built into the TMS 32010(M) architecture to allow for a smooth and relatively quick progression to the next microprocessors and microcomputers in the family.

5.2 320 Description

The TMS 320 family's TMS 32010 and TMS 320M10 microprocessor and microcomputer are based on a modified Harvard architecture. The definition of Harvard structures separates program and data spaces enabling complete overlap of instruction fetch and execution. The modification implemented in these devices allows bidirectional transfers between the two spaces via a bus interchange module (BIM). The increased flexibility that this gives permits contents stored in

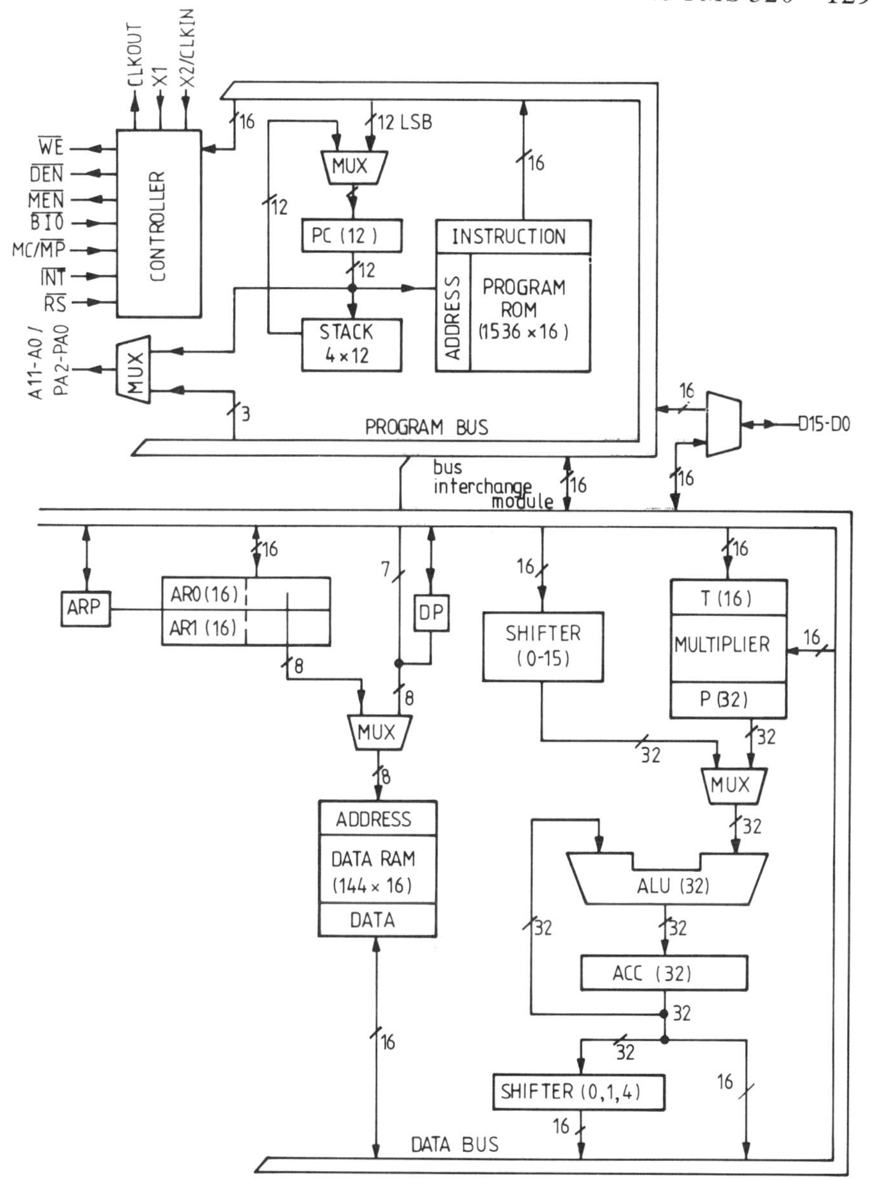

Figure 5.1 TMS 320M10 functional block diagram

program memory to be passed into data space. This mechanism obviates the need for a tri-space architecture requiring a fixed size coefficient ROM in addition to program and data memories. The BIM provides for computed subroutine calls as well as immediate instructions.

A block diagram of the TMS 320M10 layout is shown in fig. 5.1. The TMS 32010 microprocessor is identical except for the omission of

Plate 5.1 TMS 32010 bar photograph

the program ROM. The first release of the parts are manufactured with a 3μ NMOS process. This technology has been refined at Texas Instruments over several general purpose microprocessor developments.

The initial devices offer a 200 ns cycle time (5 MHz) derived from an external 20 MHz crystal or clock. A large percentage of the instructions are single words, single cycle and are described in section 5.3. An instruction prefetch mechanism is implemented to increase throughput.

5.2.1 Memory Spaces

The program memory address space is 4K $\times$ 16-bit words. 1.5K $\times$ 16 of this can be mask programmed into the program ROM on the TMS 320M10. The additional 2.5K $\times$ 16-bit can be expanded off-chip. At any time the on chip program memory can be switched off by the MC/$\overline{\text{MP}}$ pin allowing complete emulation of the TMS 32010 microprocessor, i.e. full 4K $\times$ 16 off-chip access. Provided that memories with access times commensurate with machine cycle time

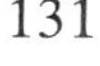

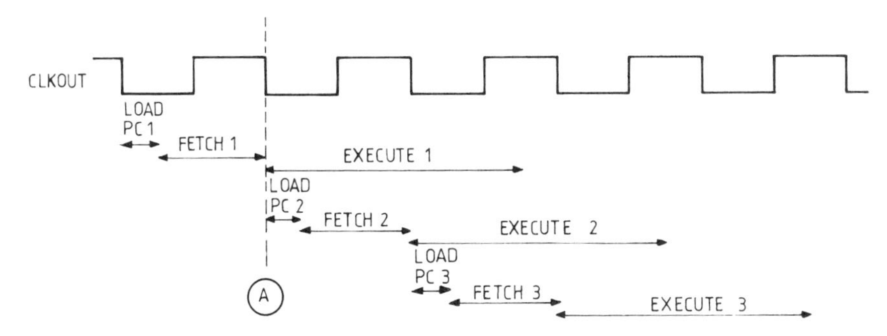

Figure 5.2 Instruction pipelining

and overheads are used, external accesses are executed at full speed. For a 200 ns instruction cycle, memories with access times of less than 100 ns are required.

The data memory comprises 144 × 16-bit RAM on-chip. All operands for instruction execution are fetched from this space. Although direct off-chip access of operands is not supported, two mechanisms are provided to bring data into the data RAM. The first allows input from a peripheral via an IN instruction. The second flexes the BIM via a TBLR (table look-up and read) instruction to transfer data from program memory. The OUT and TBLW instructions perform complementary operations in writing from data RAM to a peripheral or program memory (if off-chip RAM).

External off-chip memory accesses use the non-multiplexed 16-bit bidirectional data bus D0-D15 and 12-bit address bus A0-A11.

The highly pipelined operation of the TMS 32010(M) is highlighted in fig. 5.2. Instruction fetches and executions are referenced to CLKOUT. On falling edge 'A' the program counter is loaded (LOAD PC 2), and the instruction is prefetched (FETCH 2) whilst execution of the previous instruction (EXECUTE 1) is still operative. A further prefetch (FETCH 3) is shown whilst both EXECUTE 1 and 2 operations are still active.

5.2.2 Arithmetic Logic

Computation on the TMS 32010(M) is based on two's complement fixed point arithmetic. The main sites of execution are: the ALU, the accumulator, the multiplier and the shifters.

The ALU operates on 32-bit data words provided by the multiplier and data memory via the 0–15-bit barrel shifter. The accumulator provides the primary operand and is the only destination register. A special feature to handle arithmetic overflows has been added. A software selectable option causes the accumulator to be loaded with the most positive or negative value of the ALU, should an overflow occur. This gives a digital equivalent of the analog 'saturation' condition.

The 32-bit accumulator is referenced by the high order and lower order words. For example storage of the two words back in data memory is achieved via SACH (store high) and SACL (store low) instructions. A set of seven branch instructions dependent on the accumulator status is supported. Should an overflow occur the event is latched in the OV register. Action such as branch on overflow can therefore be taken at some time after the event, e.g. outside of time critical code, following examination of the OV register.

The 16 × 16-bit multiplier is shown (fig. 5.1) comprising three parts: the temporary 16-bit T register, the hardware multiplier and a 32-bit product P register. To perform the multiplication function, loading of the T register via one of the three instructions is followed by either a multiply immediate instruction (providing a 13-bit value embedded in the instruction word) or a multiply instruction (providing a 16-bit value addressed in the data memory). The product is placed in the P register which can be added, subtracted or directly loaded into the accumulator.

As the P register can not be directly restored hardware prevents an interrupt occurring until one instruction after a multiplication. This allows the contents of the P register to be stored prior to any interrupt service routine.

The continuous multiply/accumulate function often required in DSP algorithms can be achieved in 400 ns by using two instructions, one from the load T register set and one from the multiply set.

The 0–15 bit barrel shifter provides for an arithmetic left shift, by zero filling of the lower order end and sign extending at the high order end. Arithmetic shift right and logical shift right, where the MSBs are also zero-filled, can be realised by small routines. The sign extension of the arithmetic shift left can be suppressed allowing the 16-bit value entering the shifter to be preserved.

The parallel shifter situated between the accumulator and the data bus only affects the high order word when the accumulator contents are stored. It provides for a left shift of 0, 1 or 4. The result of a 15-bit + sign multiplied by a 15-bit + sign is a 30-bit + sign number. The most significant bit of the 32-bit accumulator is not used. The left shift of 1 removes this MSB value before the high order accumulator word is stored. When doing a 13 × 16 multiply as in the multiply immediate case the extra 4 MSB bits are in excess and are removed by the left shift of 4, repositioning the 16 MSBs of the value in the high order position.

The following example demonstrates the shifter's operation.

Accumulator

MSB 31	15	0 LSB
F 3 2 C	7 6 3 A	

Following the instruction SACH 60, 4 (store high order accumulator word in location 60 following a left shift of 4) the data in location 60 will be 32C7.

These shifts do not affect the accumulator. The original value is still held.

5.2.3 Addressing Modes

Direct, indirect and immediate addressing modes are supported for program and data memory access. A total of 8 modes are available.

5.2.3.1 Indirect data memory addressing

This is achieved via one of the two auxiliary 16-bit registers AR0 and AR1. The least significant 8 bits of the registers address the data RAM. The auxiliary register pointer (ARP), a bit within the status register, determines which register is to be used. Using the auto-increment or autodecrement option in this mode allows transparent access to data tables without further direct reference.

5.2.3.2 Direct data memory addressing

Unlike the indirect case the data page pointer (DP) is used to access either page 0 containing 128×16 words or page 1 that contains 16×16 words. As shown in fig. 5.1, 7 bits contained within the instruction word on the program bus are used in conjunction with DP to give the direct addressing function.

5.2.3.3 Indirect program memory addressing

This is a very useful mode that allows the 12-bit program counter to be loaded from the accumulator. It allows calculated values to be used for program memory addressing. The CALA instruction uses this to call a subroutine situated at the address given by the twelve LSBs of the accumulator. The table look-up instructions also make use of this.

5.2.3.4 Direct program memory addressing

The two word branch and call instruction comprises a first word opcode and second word branch address.

Note Under normal conditions the program counter (PC) is automatically incremented to the next code in sequence.

5.2.3.5 Immediate addressing

Five instructions allow immediate operands to be embedded within the opcode. An example of this is the MPYK (multiply immediate) which contains a 13-bit value to be multiplied with the T register.

5.2.4 Registers

5.2.4.1 Status register

The status register consists of 5-bits. The contents of the register can be saved in data memory using the SST (store status) instruction. It can be altered or reloaded via the LST (load status) instruction. The exception is that the interrupt mask bit (INTM) can only be affected by the interrupt enable and disable instructions EINT and DINT.

Interrupt mask bit (INTM)

This bit shows whether the interrupt is either enabled or disabled. On servicing an interrupt, INTM is automatically set reflecting a disabling of the interrupt. It can be reset on completion of the service routine.

Data page point (DP)

Used in the direct addressing mode to access page 0 or page 1 of data memory. Two dedicated instructions as well as LST can change its value.

Auxiliary register pointer (ARP)

Used to specify either AR0 or AR1 for indirect addressing.

Overflow mode bit (OVM)

Enables/disables the overflow options (see section 5.2).

Overflow flag register (OV)

Indicates whether or not the accumulator has overflowed.

When storing the status register in the direct addressing mode it is always loaded into page 1 of the data memory in a location 0 to 15 specified by the direct address.

5.2.4.2 Auxiliary registers

In addition to the indirect addressing of data memory the two auxiliary registers can be used for loop control and temporary storage.

The BANZ (branch on auxiliary register zero) checks the least significant nine bits of the register in use. If these are not zero the register is decremented and a branch occurs. The following example (program 5.1) demonstrates this loop control mechanism.

This routine causes the values in data memory 6 through 0 to be summed with the accumulator.

```
        LARP AR1     Load ARP with 1.
                     (selecting auxiliary register 1)

        LARK AR1,6   Load AR1 with 6

LOOP    ADD  *       Add data memory to accumulator
                     (indirect mode)

        BANZ LOOP    Branch if not zero to LOOP
```

Program 5.1 Loop control with auxiliary registers

5.2.5 Interrupts

The $\overline{\text{INT}}$ pin (fig. 5.5.) provides for a hardware vectored interrupt. On acknowledging the interrupt, address value 2 is placed in the PC and the interrupt is serviced from this location or from the branch address specified in location 3 following the branch opcode in location 2 (see fig. 5.3).

The $\overline{\text{RS}}$ pin allows the TMS 32010(M) to be reset by putting the machine into a known state with value zero placed in the PC (see fig. 5.3).

The $\overline{\text{BIO}}$ pin is often referred to as the software interrupt line. This caters for bit test and branch. If in the low state when a BIOZ instruction is executed a branch occurs.

5.2.6 Program Memory Support

The 12-bit program counter (PC) is used for program memory addressing and is buffered onto the address lines A0–A11. Lines A0–A2 are multiplexed with the input/output port address lines.

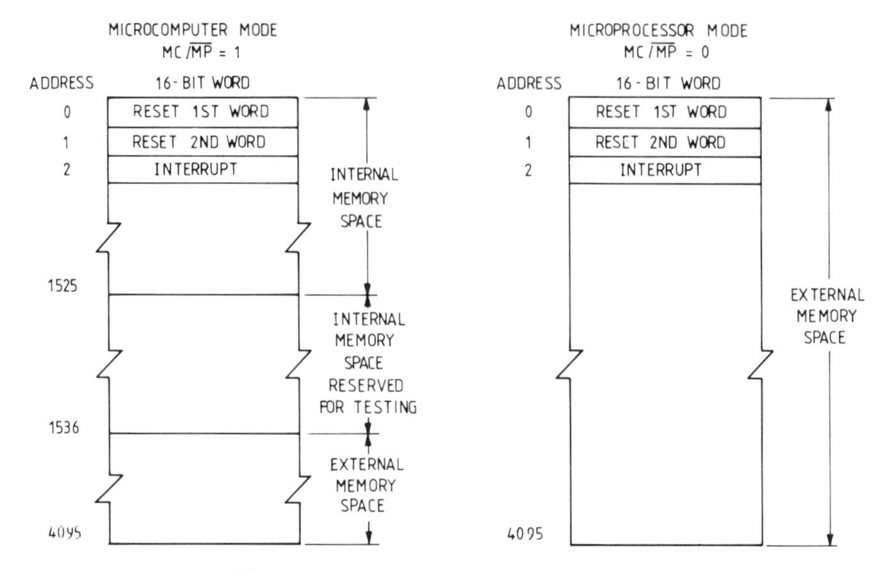

Figure 5.3 TMS 32010(M) memory maps

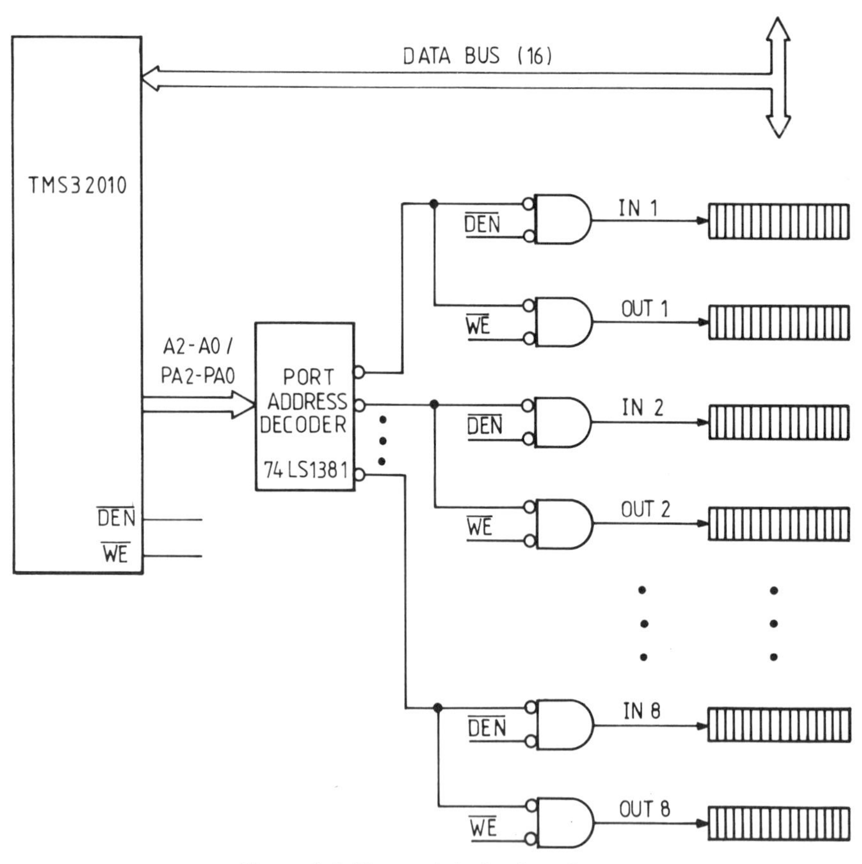

Figure 5.4 External device interface

A 4 level stack is implemented. This can be expanded via the POP and SACL instruction to further embed stack layers in data RAM and should it be needed, via the TBLW instruction, extremely large stacks can be structured in off-chip memory. The POP instruction, to remove the top of stack, is complemented by a PUSH instruction placing the 12 LSBs of the accumulator onto the stack.

5.2.7 Input/output

In a similar manner to many general purpose single microcomputers, input and output to non memory mapped peripherals is supported. Eight 16-bit input and eight 16-bit output ports give a maximum transfer rate of 2.5 Mbits/second. Two double word instructions IN and OUT support this function. Fig. 5.4 shows a simple peripheral interface system.

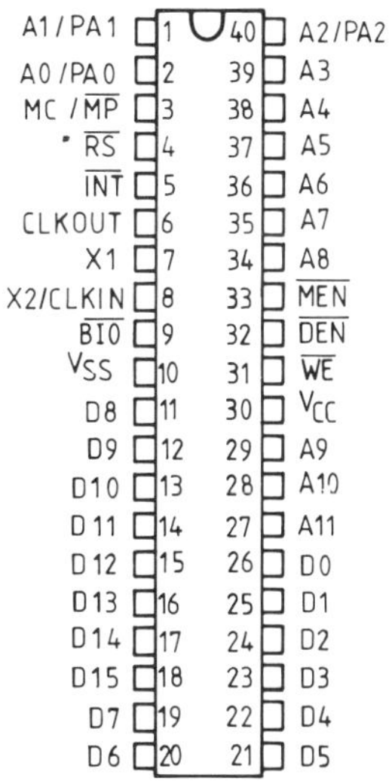

Signal	Pin	Pin	Signal
A1/PA1	1	40	A2/PA2
A0/PA0	2	39	A3
MC/$\overline{\text{MP}}$	3	38	A4
$\overline{\text{RS}}$	4	37	A5
$\overline{\text{INT}}$	5	36	A6
CLKOUT	6	35	A7
X1	7	34	A8
X2/CLKIN	8	33	$\overline{\text{MEN}}$
$\overline{\text{BIO}}$	9	32	$\overline{\text{DEN}}$
V_{SS}	10	31	$\overline{\text{WE}}$
D8	11	30	V_{CC}
D9	12	29	A9
D10	13	28	A10
D11	14	27	A11
D12	15	26	D0
D13	16	25	D1
D14	17	24	D2
D15	18	23	D3
D7	19	22	D4
D6	20	21	D5

Signature	**I/O**	**Definition**
A11-A0/ PA2-PA0	**OUT**	**External address bus. I/O port address multiplexed over PA2-PA0.**
$\overline{\text{BIO}}$	**IN**	**External polling input for bit test and jump operations.**
CLKOUT	**OUT**	**System clock output, $\frac{1}{4}$ crystal/ CLKIN frequency.**
D15-D0	**I/O**	**16-bit data bus.**
$\overline{\text{DEN}}$	**OUT**	**Data enable indicates the processor accepting input data on D15-D0.**
$\overline{\text{INT}}$	**IN**	**Interrupt.**
MC/$\overline{\text{MP}}$	**IN**	**Memory mode select pin. High selects microcomputer mode. Low selects microprocessor mode.**
$\overline{\text{MEN}}$	**OUT**	**Memory enable indicates that D15-D0 will accept external memory instruction.**
$\overline{\text{RS}}$	**IN**	**Reset used to initialize the device.**
V_{CC}	**IN**	**Power.**
V_{SS}	**IN**	**Ground.**
$\overline{\text{WE}}$	**OUT**	**Write enable indicates valid data on D15-D0.**
X1	**In**	**Crystal input.**
X2/CLKIN	**IN**	**Crystal input or external clock input.**

Figure 5.5 TMS 32010(M) pin out and description

5.2.8 Pin Out

The commercial temperature devices are available in 40 pin ceramic or plastic D.I.L. packages. Fig. 5.5 shows the pin out and description.

5.3 320 Instruction Set

The general purpose high numeric capability of the TMS 32010 as well as its specific application to digital signal processing problems is well highlighted by the instruction set. Table 5.2 lists the 60 instructions comprising in the main single word single cycle operations. The shift operations performed by the two shifters are an optional extra when carrying out certain arithmetic functions and no extra time penalty is paid for their use.

Table 5.2 is divided into six sections dealing with the subsets of

Table 5.1 Instruction symbols

Symbol	Meaning
ACC	Accumulator
AR0	Auxiliary register zero
AR1	Auxiliary register one
	(AR0 and AR1 are predefined assembler symbols equal to 0 and 1, respectively.)
ARP	Auxiliary register pointer
D	Data memory address field
DATn	Label assigned to data memory location n
dma	Data memory address
DP	Data page pointer
I	Addressing mode bit
K	Immediate operand field
(N)	Contents of register "N" or data memory location "N"
>nn	Indicates nn is a hexadecimal number. All others are assumed to be decimal values.
P	Product (P) register
PAn	Port address n (PA0 through PA7 are predefined assembler symbols equal to 0 through 7, respectively)
PC	Program counter
pma	Program memory address
PRGn	Label assigned to program memory location n
R	1-bit operand field specifying auxiliary register
S	4-bit left-shift code
T	T register
X	3-bit accumulator left-shift field
→	Is assigned to

Table 5.2 Instruction set summary

Mnemonic	Description	No. Cycles	No. Words	15	14	13	12	11	10	9	8	7	6	5	4	3	2	1	0
				Opcode Instruction Register															
	Accumulator Instructions																		
ABS	Absolute value of accumulator	1	1	0	1	1	1	1	1	1	1	1	0	0	0	1	0	0	0
ADD	Add to accumulator with shift	1	1	0	0	0	0	←	S		→	I	←			D			→
ADDH	Add to high-order accumulator bits	1	1	0	1	1	0	0	0	0	0	I	←			D			→
ADDS	Add to accumulator with no sign extension	1	1	0	1	1	0	0	0	0	1	I	←			D			→
AND	AND with accumulator	1	1	0	1	1	1	1	0	0	1	I	←			D			→
LAC	Load accumulator with shift	1	1	0	0	1	0	←	S		→	I	←			D			→
LACK	Load accumulator immediate	1	1	0	1	1	1	1	1	1	0	←				K			→
OR	OR with accumulator	1	1	0	1	1	1	1	0	1	0	I	←			D			→
SACH	Store high-order accumulator bits with shift	1	1	0	1	0	1	1	←	X	→	I	←			D			→
SACL	Store low-order accumulator bits	1	1	0	1	0	1	0	0	0	0	I	←			D			→
SUB	Subtract from accumulator with shift	1	1	0	0	0	1	←	S		→	I	←			D			→
SUBC	Conditional subtract (for divide)	1	1	0	1	1	0	0	1	0	0	I	←			D			→
SUBH	Subtract from high-order accumulator bits	1	1	0	1	1	0	0	0	1	0	I	←			D			→
SUBS	Subtract from accumulator with no sign extension	1	1	0	1	1	0	0	0	1	1	I	←			D			→
XOR	Exclusive OR with accumulator	1	1	0	1	1	1	1	0	0	0	I	←			D			→
ZAC	Zero accumulator	1	1	0	1	1	1	1	1	1	1	1	0	0	0	1	0	0	1

Table 5.2 cont'd

Mnemonic	Description	No. Cycles	No. Words	Opcode Instruction Register 15	14	13	12	11	10	9	8	7	6	5	4	3	2	1	0
	Accumulator Instructions (cont'd)																		
ZALH	Zero accumulator and load high-order bits	1	1	0	1	1	0	0	1	0	1	I	←			D			→
ZALS	Zero accumulator and load low-order bits with no sign extension	1	1	0	1	1	0	0	1	1	0	I	←			D			→
	T Register, P Register, and Multiply Instructions																		
APAC	Add P register to accumulator	1	1	0	1	1	1	1	1	1	1	1	0	0	0	1	1	1	1
LT	Load T register	1	1	0	1	1	0	1	0	1	0	I	←			D			→
LTA	LTA combines LT and APAC into one instruction	1	1	0	1	1	0	1	1	0	0	I	←			D			→
LTD	LTD combines LT, APAC, and DMOV into one instruction	1	1	0	1	1	0	1	0	1	1	I	←			D			→
MPY	Multiply with T register; store product in P register	1	1	0	1	1	0	1	1	0	1	I	←			D			→
MPYK	Multiply T register with immediate operand; store product in P register	1	1	1	0	0	←							K			→
PAC	Load accumulator from P register	1	1	0	1	1	1	1	1	1	1	1	0	0	0	1	1	1	0
SPAC	Subtract P register from accumulator	1	1	0	1	1	1	1	1	1	1	1	0	0	1	0	0	0	0

Control Instructions																			
DINT	Disable interrupt	1	1	0	1	1	1	1	1	1	1	1	0	0	0	0	0	0	1
EINT	Enable interrupt	1	1	0	1	1	1	1	1	1	1	1	0	0	0	0	0	1	0
LST	Load status register	1	1	0	1	1	1	1	0	1	1	I	←			D			→
NOP	No operation	1	1	0	1	1	1	1	1	1	1	1	0	0	0	0	0	0	0
POP	Pop stack to accumulator	2	1	0	1	1	1	1	1	1	1	1	0	0	1	1	1	0	1
PUSH	Push stack from accumulator	2	1	1	1	1	1	1	1	1	1	1	0	0	1	1	1	0	0
ROVM	Reset overflow mode	1	1	0	1	1	1	1	1	1	1	1	0	0	0	1	0	1	0
SOVM	Set overflow mode	1	1	0	1	1	1	1	1	1	1	1	0	0	0	1	0	1	1
SST	Store status register	1	1	0	1	1	1	1	1	0	0	I	←			D			→
I/O and Data Memory Operations																			
DMOV	Copy contents of data memory location into next location	1	1	0	1	1	0	1	0	0	1	I	←			D			→
IN	Input data from port	2	1	0	1	0	0	0		PA		I	←			D			→
OUT	Output data to port	2	1	0	1	0	0	1		PA		I	←			D			→
TBLR	Table read from program memory to data RAM	3	1	0	1	1	0	0	1	1	1	I	←			D			→
TBLW	Table write from data RAM to program memory	3	1	0	1	1	1	1	1	0	1	I	←			D			→

Table 5.2 cont'd

Mnemonic	Description	No. Cycles	No. Words	15	14	13	12	11	10	9	8	7	6	5	4	3	2	1	0
				Opcode Instruction Register															
Auxiliary Register and Data Page Pointer Instructions																			
LAR	Load auxiliary register	1	1	0	0	1	1	1	0	0	R	I	←			D			→
LARK	Load auxiliary register immediate	1	1	0	1	1	1	0	0	0	R	←				K			→
LARP	Load auxiliary register pointer immediate	1	1	0	1	1	0	1	0	0	0	1	0	0	0	0	0	0	K
LDP	Load data memory page pointer	1	1	0	1	1	0	1	1	1	1	I	←			D			→
LDPK	Load data memory page pointer immediate	1	1	0	1	1	0	1	1	1	0	0	0	0	0	0	0	0	K
MAR	Modify auxiliary register and pointer	1	1	0	1	1	0	1	0	0	0	I	←			D			→
SAR	Store auxiliary register	1	1	0	0	1	1	0	0	0	R	I	←			D			→

Branch Instructions				
B	Branch unconditionally	2	2	1 1 1 1 1 0 0 1 0 0 0 0 0 0 0 0
				0 0 0 0 ← Branch Address →
BANZ	Branch on auxiliary register not zero	2	2	1 1 1 1 0 1 0 0 0 0 0 0 0 0 0 0
				0 0 0 0 ← Branch Address →
BGEZ	Branch if accumulator $\geqslant 0$	2	2	1 1 1 1 1 1 0 1 0 0 0 0 0 0 0 0
				0 0 0 0 ← Branch Address →
BGZ	Branch if accumulator > 0	2	2	1 1 1 1 1 1 0 0 0 0 0 0 0 0 0 0
				0 0 0 0 ← Branch Address →
BIOZ	Branch on $\overline{BIO} = 0$	2	2	1 1 1 1 0 1 1 0 0 0 0 0 0 0 0 0
				0 0 0 0 ← Branch Address →
BLEZ	Branch if accumulator $\leqslant 0$	2	2	1 1 1 1 1 0 1 1 0 0 0 0 0 0 0 0
				0 0 0 0 ← Branch Address →
BLZ	Branch if accumulator < 0	2	2	1 1 1 1 1 0 1 0 0 0 0 0 0 0 0 0
				0 0 0 0 ← Branch Address →
BNZ	Branch if accumulator $\neq 0$	2	2	1 1 1 1 1 1 1 0 0 0 0 0 0 0 0 0
				0 0 0 0 ← Branch Address →
BV	Branch on overflow	2	2	1 1 1 1 0 1 0 1 0 0 0 0 0 0 0 0
				0 0 0 0 ← Branch Address →
BZ	Branch if accumulator $= 0$	2	2	1 1 1 1 1 1 1 1 0 0 0 0 0 0 0 0
				0 0 0 0 ← Branch Address →
CALA	Call subroutine from accumulator	2	1	0 1 1 1 1 1 1 1 1 0 0 0 1 1 0 0
CALL	Call subroutine immediately	2	2	1 1 1 1 1 0 0 0 0 0 0 0 0 0 0 0
				0 0 0 0 ← Branch Address →
RET	Return from subroutine	2	1	0 1 1 1 1 1 1 1 1 0 0 0 1 1 0 1

accumulator, auxiliary register and data page pointer, branch, multiply associated, control and I/O and data memory instructions.

Certain of the instructions are familiar and straightforward to those with even limited understanding of microprocessor machines. Due to their potential power a few instructions are discussed in more detail.

5.3.1 Accumulator Instructions

The instructions ADDH, ADDS, SUBH, SUBS, ZALS can be used to implement 32-bit arithmetic with the TMS 320. In the following example (program 5.2) a 32-bit subtraction is performed between word 1 (MSW1, LSW1) and word 2 (MSW2, LSW2) made up of the data memory locations MSW1, LSW1, MSW2 and LSW2.

The results reside in the accumulator. The SUBC (conditional subtract) is the vehicle for executing a divide operation. Multiplication, although explicit within the instruction set, could be implemented by a series of shift and adds. Division can likewise be synthesised by a sequence of subtracts and shifts. The following routine (program 5.3) performs this function.

The data memory locations NUMERA, DENOMI, QUOT, REMAIN and TMPSGN hold the dividend, divisor, quotient, remainder and the sign of the quotient, respectively.

```
*
SUB32B ZALS LSW1     Load ACC with least significant 16-bits of word 1

       ADDH MSW1     Load ACC with most significant 16-bits of word 1

       SUBS LSW2     Subtract least significant 16-bits of word 2 from 1

       SUBH MSW2     Subtract most significant 16-bits of word 2 from 1
```

Program 5.2 32-bit arithmetic

5.3.2 Auxiliary Register Instructions

The LAR instruction (load auxiliary register) and its complementary instruction SAR (store auxiliary register) should be used to store and load the auxiliary registers during subroutine calls and interrupts. If, however, an auxiliary register is not being used for indirect addressing, the LAR and SAR instructions enable it to be used as an additional storage register. It is especially useful for swapping values between data memory locations. When using the LAR instruction in indirect

```
*
NUMERA    EQU >10
DENOMI    EQU >20
QUOT      EQU >30
TMPSGN    EQU >40
REMAIN    EQU >50
*
DIVI      LARP 0          Select AR0 as loop counter
          LAC NUMERA      Put dividend in ACC
          XOR DENOMI      XOR with divisor to determine size of quotient
          SACL TMPSGN     Store sign of quotient in TMPSGN
          LAC DENOMI      Make divisor positive
          ABS
          SACL DENOMI
          LAC NUMERA      Align dividend and make it positive
          ABS
          LARK 0,15       Set loop count to 15
*
*Conditional Divide Loop
*
CDLOOP    SUBC DENOMI     Conditonal subtract divisor
          BANZ CDLOOP     Repeat loop 15 times
          SACL QUOT       Store resultant quotient in QUOT
          LAC TMPSGN      Check sign of quotient
          BGEZ DONE       Done if positive
          ZAC             Invert quotient if negative
          SUB QUOT
          SACL QUOT       Store negated quotient in QUOT
*
DONE      SACH REMAIN     Store remainder in REMAIN
```

Program 5.3 Division routine

addressing with auto-decrement/auto-increment, the new value of the auxiliary register is not decremented/incremented as a result of the instruction execution. However, in contrast to the instruction LAR, the SAR instruction will perform the auto-decrement of the auxiliary register as shown below (program 5.4).

The sequence of events in the internal hardware is as follows:

(1) The current value of AR0 before execution of the LAR and SAR instructions is used to address the data memory for the loading and storing operations.

```
*
LOAD      LARP AR0        Select AR0
          LARK AR0,7      AR0 = 7 and assume
*                         data memory 7 content = 30
          LAR AR0,*-      AR0 = 30. No auto-decrement will take
                          place.
*
STORE     LARP AR0        Select AR0.
          LARK AR0,10     AR0 = 10.
          SAR AR0,*-      Puts value 9 in memory location 10.
*         SAR AR0,*+      Puts value 11 in data memory location 10
*
```

Program 5.4 Auxiliary register manipulation

(2) The indirect auto-increment/auto-decrement operation is performed.

(3) The contents of the addressed data memory is loaded into AR0 in the case of the LAR instruction. In the case of the SAR instruction, the auto-incremented/auto-decremented value of AR0 will then be stored in the addressed data memory.

5.3.3 Branch Instructions

The CALA instruction permits subroutine calls from computed addresses. The following example (program 5.5) routine reads in a 16-bit data word from an I/O device and extracts bits 0 and 1 to compute one of the subroutine addresses SUB1, SUB2, SUB3 or SUB4.

The CALA instruction causes a branch to the relevant subroutine as shown below (data memory locations MASK, VALUE, TEMP, RESULT are assumed).

5.3.4 Multiply Associated Instruction

Two of the most important instructions on the TMS 32010(M) are MPY and LTD. Together they form the essential elements of multiply accumulate and shift of a transversal structure. An example finite impulse response (FIR) filter program is shown in section 5.6.3. The LTD instruction combines the functions of LT, APAC and DMOV in one single cycle instruction. A value is loaded into the T register. The previous multiply result in the P register is accumulated and then the contents of the addressed data memory location are copied into the

```
        SUB1 EQU >A0
        SUB2 EQU >A6
        SUB3 EQU >AC
        SUB4 EQU >B2
*
LACK >A0          Initialise ACC with >A0.
SACL TEMP         Save in data memory TEMP.
LACK 3            Load ACC with 3.
SACL MASK         Set up comparison bit mask.
IN VALUE,PA3      Read input through port PA3 into VALUE.
LAC VALUE         Load ACC with content of VALUE.
AND MASK          Logical AND MASK with ACC.
SACL RESULT       Store comparison results in RESULT.
ZALS TEMP         Zero ACC, and load ACC with TEMP.
LT RESULT         Load TREG with content of RESULT.
MPYK 6            Calculate subroutine address.
APAC              Add product to ACC as call address.
CALA              Call computed routine.
```

Program 5.5 Computed calls

next highest location. The MPY instruction will multiply the T register by a 16-bit value and place the result in the P register. These two instructions will very often be found placed alternately in a long string of code representing high order filters.

5.3.5 I/O and Data Memory Operations

TBLR and TBLW, the table look up instructions, are the only tri-cycle instructions in the set. The TBLR perform the following operations:

(i) Contents of PC + 1 are pushed onto the top of stack (TOS).
(ii) The contents of the LS 12 bits of ACC are loaded into the PC.
(iii) The contents on program memory pointed to by PC are addressed by A0–A11 and transferred via D0–D15 into the specified data memory location.
(iv) The TOS is popped back into the PC.

TBLW perform the inverse function in writing from data memory into program memory.

5.4 320 Development Process

The TMS 320 family concept has been extended to the hardware and software development products. The concept of XDS (extended development support) has been implemented on other Texas Instruments processor families, but the first complete package was offered for the TMS 320. XDS provides tools from initial evaluation to final hardware test and debug. The strategy of XDS is to provide host independent support, i.e. development software for all major development host machines and hardware linked via non dedicated bus structures to allow use with many hosts and terminals.

5.4.1 XDS/320 Evaluation Module

The evaluation module (EVM) provides a stand alone single board sporting the functionality to allow comprehensive evaluation of the TMS 320 for a particular application. Two EIA ports can be used for connection to a host computer, terminal or a line printer. In addition to accepting object code the EVM can also accept source code to be assembled by the on board symbolic assembler. The monitor firmware also includes a reverse assembler, a very useful patch assembler and a line-oriented text editor.

The debug monitor allows single stepping program memory, breakpoints and access to all of the TMS 32010's registers.

Plate 5.2 XDS/320 emulator connected to the XDS/320 analog interface board

A read/write audio cassette interface for code storage is provided along with an EPROM programmer with all voltages generated from the board supply. A target connector supplied can be used to effectively emulate the TMS 32010 in a target system. This cable is used to connect the EVM with the analog interface board (AIB), as shown in plate 5.2.

5.4.2 XDS/320 Analog Interface Board

Invariably the TMS 32010 is evaluated for use in processing analog signals. A/D conversion is required and the results of processing are required in the analog domain by D/A conversion. The analog interface board (AIB) provides for this evaluation. The board has a 12-bit A to D and a 12-bit D to A converter. The A to D gives a maximum conversion time of 25 μs with an analog input of +/− 10 volts. The sampling rate is programmable between 15 Hz and 15 kHz (15 Hz resolution).

Low pass filters for antialiasing and output smoothing are programmable with a bandwidth of 50 Hz to 4 kHz. Digital inputs and outputs allow external converters to be given access to the TMS 32010 EVM processing. An audio amplifier is provided on the analog output that can be connected to an 8 ohm speaker.

The AIB provides access to the EVM for the majority of signals within the bandwidth of TMS 32010 processing.

5.4.3 XDS/320 Macro Assembler/Linker

This software package translates TMS 32010 source assembler language into executable object code. In addition to the standard instruction set, mnemonics representing a suite of macros provided by the package are supported. A listing and object file with optional symbol and cross-reference listing are produced by the package. The linker allows application code developed on a modular basis and library routines to be linked together.

The concept of XDS is to be host independent. The macro assembler/linker is available for a variety of host and operating systems including IBM 370 (MVS, CMS), VAX (VMS) and TI990.

5.4.4 XDS/320 Simulator

This software program provides full simulation of the TMS 32010(M) allowing quick application program check out. The simulator has a debug option similar to that of the hardware emulator. The state of the machine can be monitored instruction by instruction.

Object code provided by the XDS/320 macro assembler is used by the simulator, and I/O devices can be simulated to give a system context to the operation. A full record of the execution, breakpointing and tracing can be kept by a debug log included.

5.4.5 XDS/320 Emulator

The emulator has become known as the 'XDS'. The devices provide the tools for full real time in-circuit emulation of a TMS 32010(M) target system. Two EIA ports support connection to a host computer, terminal, PROM programmer or printer. Via one port object, from a host, created by the XDS/320 assembler linker can be downloaded into the emulator. The second port can provide system control via a terminal. The unit allows for target or emulator clocks and memories to be selected. The debug capabilities include both software and hardware breakpoints. A total of ten software breakpoints can halt emulation when specific instruction acquisitions occur. The two hardware breakpoints allow halting on more complex events, e.g. on any memory cycle, any two addresses, specified data value or opcode block.

The emulator provides a trace function that can be viewed as a 'cine camera' recording a sequence of events during the emulation session. This allows reviewing of the machine status up to the breakpoint.

The emulator provides a flying lead with a 32010(M) target connector. In addition another lead with TTL compatible probes attached allows external logic states to be used as qualifiers and triggers. The emulator design provides for debug of complex multiprocessing systems. Up to nine emulators can be daisy chained. They can be individually controlled or fully synchronised. The emulators need not be all TMS 32010 versions. The XDS concept supports other Texas Instruments processor families. Hence it is possible to have a system comprising a TMS 9995, TMS 99000, TMS 7000 and several TMS 32010s being completely debugged using this capability.

The emulator will be offered with a high level debug procedural language similar to TI AMPL and possibly including local mass storage by the addition of controller cards to the chassis.

In the short time since the TMS 320 launch a range of third party products supporting it have emerged. Assemblers and simulators are available for CP/M, MSDOS, PDP11 and HP machines, along with design packages, e.g. filter design utilities.

5.5 General Application Techniques

The following sections deal with some specific application studies. Presented here are some more general system design ideas and methods using the TMS 32010(M).

5.5.1 Interprocessor Communication

Fig. 5.6 shows a 16-bit dual-ported register to allow data to flow bidirectionally between the TMS 32010 and a 16-bit processor. The diagram shows the Input/Output port address lines PA0–PA2 being decoded to provide a handshake mechanism via the interrupt feature. The use of the software interrupt $\overline{\text{BIO}}$ for synchronisation from the processor acknowledge line is shown.

The $\overline{\text{BIO}}$ pin can be used to monitor the status of a peripheral device serviced by the TMS 32010. This method of polling has the distinct advantage that fixed critical software loops can be executed without interruption by external hardware. Moreover, since the device status/interrupt is polled at defined points in the software loops, context save is minimal. The following example, routine (program 5.6) assumes that the $\overline{\text{BIO}}$ pin is connected to an external FIFO which contains sixteen data words. This FIFO is only serviced when full and the $\overline{\text{BIO}}$ pin is polled after each time-critical function has been executed.

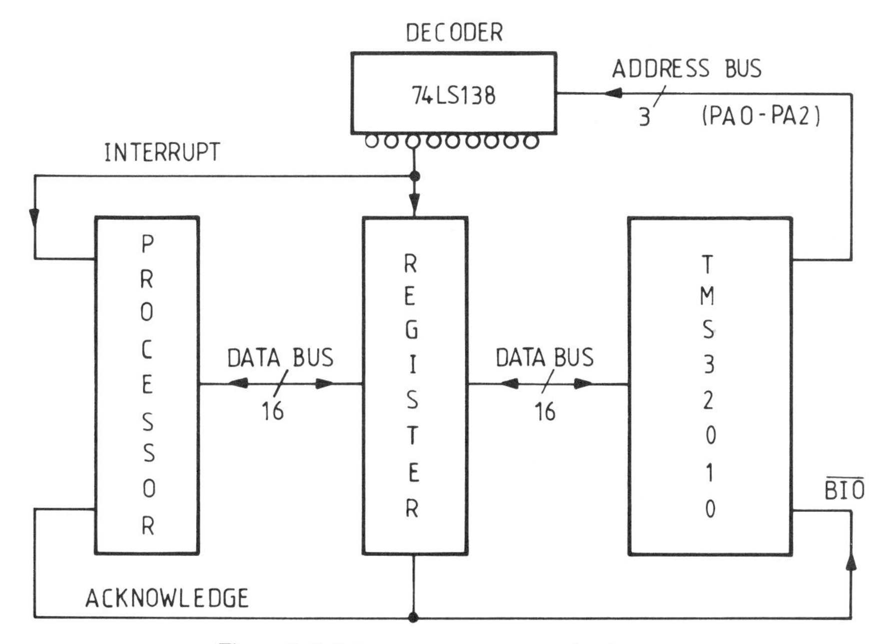

Figure 5.6 Interprocessor communication

```
*
* Main Program Segment
*
POLL      BIOZ SKIP          Poll device status, if signal low
                             then skip.
          CALL SERVE         otherwise, call device service
                             routine.
SKIP      *                  Continue with user's time-critical
                             loop.
          *
*
* Device Service Routine
*
SERVE     LARK AR0,15        Set up count of 16 in auxiliary
                             register AR0.
          LARK AR1,TABLE     Set up data memory address for
                             transfer.
LOOP      LARP 1             Point to auxiliary register AR1 for
                             transfer.
          IN PA0,*+AR0       Read input via PA0. Increment
                             AR1. Select AR0.
          BANZ LOOP          Loop and decrement AR0 if not equal
                             to zero.
          RET                If AR0 is zero then return to
                             user's program.
*
```

Program 5.6 Interprocessor communication

This device service routine does not have to save the status and contents of the internal registers since a new procedure will be executed after the FIFO has been serviced.

Another example routine (program 5.7) below shows how the TMS 32010 fills a FIFO of unknown depth from its data memory. Again the FIFO-full signal (logic one) is connected to the $\overline{\text{BIO}}$ pin.

```
INIT      LARP 0         Load ARP with zero to select AR0.
          LARK AR0,0     Initialise AR0 with zero.
LDFIFO    OUT *+,PA0     Output data via PA0 and increment AR0.
          BIOZ LDFIFO    Loop load FIFO if device status is low.
```

Program 5.7 FIFO filling

5.5.2 Extended Data Memory Techniques

The facility of data memory expansion using program memory space and table read and write instructions is described in 5.2.1. However,

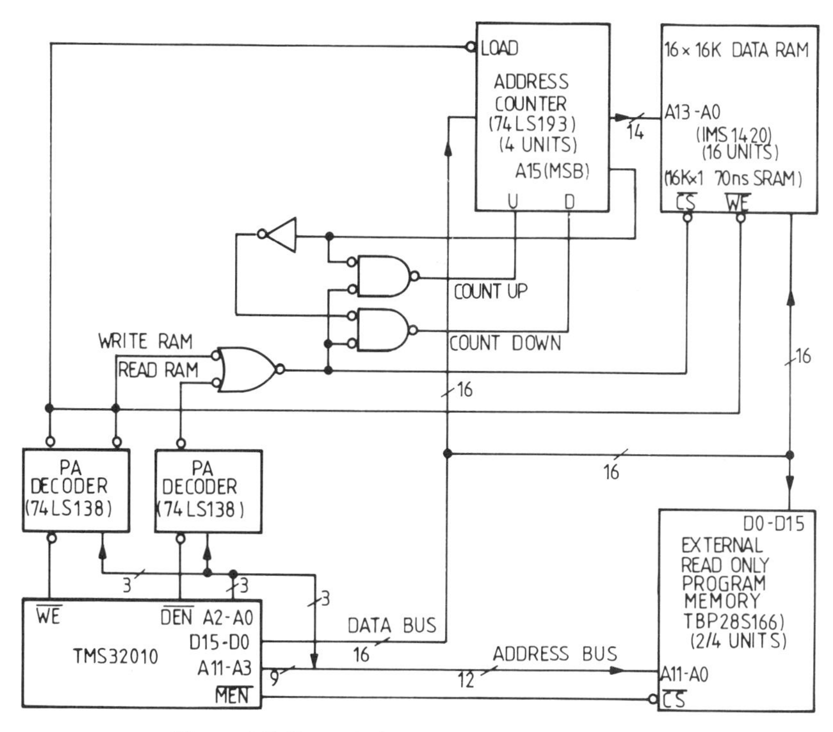

Figure 5.7 Extended data memory technique

for systems requiring more than a few kilowords of data memory and where the three cycle instruction is too time consuming, this technique is not applicable. Fig. 5.7 shows a system comprising 16 × 16K words of off-chip data memory. The address counter is seen as a peripheral I/O device to the TMS 32010 that in turn provides the address to the bank of static RAM. Access to this data could be achieved by a sequence of OUT (output to address counter) followed by IN (input of the data) instructions. However very often a block data transfer is needed. In this case the overhead of the OUT cycles can be removed as an automatic up/down counter mechanism is used. The MSB of the address counter triggers the up/down logic. Data will be read sequentially until a new value is loaded into the address counter.

A similar technique is used in the Texas Instruments professional computer TMS 32010 speech option board. The large data space is required for template storage for the word recognition facility.

5.5.3 Hardware Interrupt Handling

The following routine (program 5.8) illustrates how the complete user's context (status and internal registers of the TMS 320) can be

```
*
INTHLR   SST STATUS      Save status register in data memory
                         STATUS.
         SACL ACCH       Save ACC/PREG higher word in data memory
                         ACCH.
         SACL ACCL       Save ACC/PREG lower word in data memory
                         ACCL.
         SAR AR0,AR00    Save auxiliary register AR0 in data
                         memory AR00.
         SAR AR1,AR01    Save auxiliary register AR1 in data
                         memory AR01.
         MPYK ONE        Multiply the T-register content by
                         value 1.
         SACL TREG       Save value of T-register in data
                         memory TREG.
*
RESTOR   LT TREG         Restore original T-register value from
                         TREG.
         LAR AR0,AR00    Restore auxiliary register AR0 value
                         from AR00.
         LAR AR1,AR01    Restore auxiliary register AR1 value
                         from AR01.
         ZALH ACCH       Restore original ACC/PREG higher word
                         from ACCH.
         ADDS ACCL       Restore original ACC/PREG lower word
                         from ACCL.
         LST STATUS      Restore original status register
                         from STATUS.
         EINT            Re-enable the interrupt mask before
                         exit.
         RET             Return to previous user's environment.
```

Program 5.8 Interrupt handling

saved and restored when an interrupt occurs. Note that there is no provision to save the contents of the product register (PREG) during interrupts. Therefore both multiply instructions (MPY, MPYK) should be followed immediately by one of these instructions: PAC, APAC, SPAC, LTA, or LTD. Provision is made in the TMS 320 hardware to inhibit external interrupts during the execution of any multiply and the succeeding instruction.

5.6 Application to Digital Filters

5.6.1 Introduction

The two main subsections of digital filters FIR (finite impulse response) and IIR (infinite impulse response) are dealt with exhaustively by most theoretical texts. In essence the FIR can be viewed as a sampled data delay line. The IIR is a recursive filter formed from a second order biquadratic structure. IIR filters have direct analog equivalents. FIRs have the unique feature of providing linear phase response (so often required in the telecommunications environment). The basic multiply, accumulate and shift of the FIR executes in 400 ns on the TMS 32010 and the biquadratic structure in 2.2 μs. Some design considerations in using these filters are highlighted by reference to an example of a low frequency system. The TMS 32010 code for versions of the two filter types is also given.

The case study is that of a simple data acquisition system (DAS). The sensor channel outputs are of a relatively low bandwidth, e.g. thermal or pressure transducers. However, due to the likelihood of system noise, prevention of aliasing back into the signal band would require a relatively complex analog low pass antialiasing filter if processing is to be purely analog. Of course a filter per channel is required – analog filters cannot be multiplexed. High performance analog filters prove costly to a system when you consider there may be fifty or more channels. By providing a single high performance antialiasing digital filter and multiplexing the channels through it via an A/D converter, the analog antialiasing filters on each channel can be reduced to simple RC networks. The system cost can be dramatically reduced. Fig. 5.8 shows a block diagram of a TMS 32010 based DAS. The approach taken to implement this system is to complement the poor front end (RC) analog filtering by running the A/D converter at a very high sampling rate (over sampling), then to low pass filter in the TMS 32010 and to reduce the sample rate, i.e. implement a decimation filter.

An FIR filter has the attraction that computation is not needed until an output sample is required. So in the decimation filter with a high input rate and low output rate computational time is minimised with an FIR as opposed to an IIR filter. With an IIR filter the computational rate is proportional to the input sample rate. However, the FIR filter is more memory intensive. Basically the data memory required for an FIR is equal to the length of the filter.

Theoretical texts usually base superiority on computational time. In

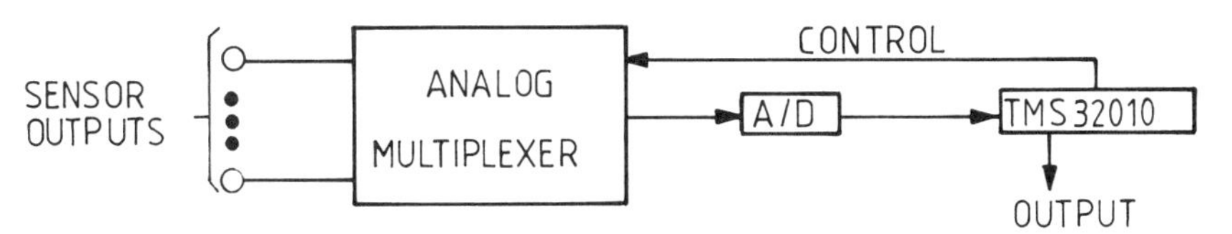

Figure 5.8 TMS 32010 based data acquisition system

a practical system factors such as memory requirements may be an overriding factor.

Quantifying the above DAS case will enable a clearer understanding of the benefits of implementing the filter digitally, and decisions involved in selecting which type, with the TMS 32010. Consider that the sensors impose the following antialiasing filter requirements:

Passband: 0 to 5 Hz (1 dB ripple)
Stopband: 50 dB attenuation at and above 10 Hz

This translates to a fourth-order elliptic filter.

In the analog domain a fourth order filter could relate to four op amps. With the associated passive components this represents an approximate 10 $ parts cost, giving probably a 30–40 $ product cost. Consider the example of 50 channels. An analog antialiasing cost of possibly 1500–2000 $ is derived. The digital implementation cost may well be 5 to 10 times smaller using the TMS 32010.

The simple block diagram of fig. 5.9 describes the DAS process and parameters. The aliasing of the analog RC frequency response is shown in fig. 5.10(a). The RC roll off gives attenuation of 34 dB at 495 Hz. A reasonable assumption that noise at the sensor output is 20 dB down is made. So if we choose an A/D conversion rate of 500 Hz the system noise aliased back into the passband will be better than 50 dB down. Fig. 5.10(b) shows the desired antialiasing frequency response of the digital low pass filter. By carrying out the decimation within the filter the effective frequency response is squashed up as shown in fig. 5.10(c). Here we have executed a 20 to 1 sample rate reduction from 500 Hz to 25 Hz. This simply means that 19 out of 20 computed terms are not used in the IIR filter or that 19 out of 20 samples are ignored in an FIR filter.

If an FIR filter is used, the sample rate reduction should not be performed by one filter. The frequency response is dependent on the stop band minus pass band difference, with respect to the sampling rate. In this case 5 Hz with a rate of 25 Hz. This system would in fact

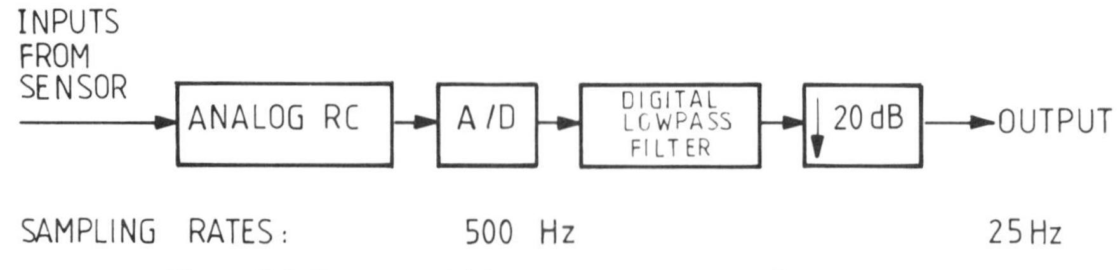

Figure 5.9 Data acquisition system process and parameters

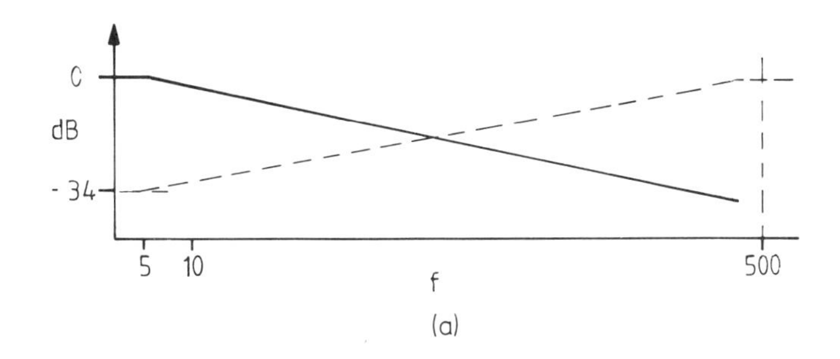

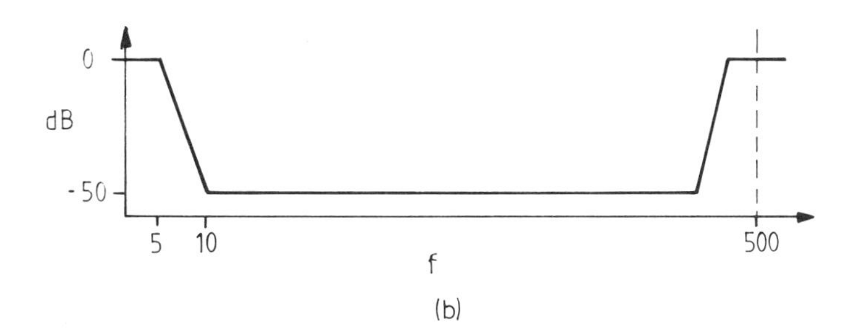

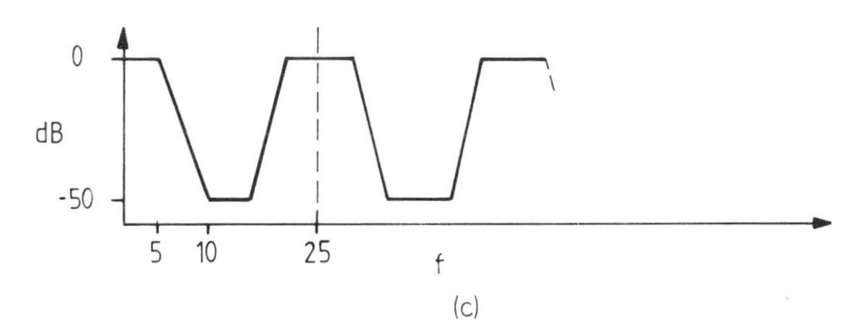

Figure 5.10 Frequency responses of analog RC and digital decimation

require a 140 tap filter. The 20 × sample rate reduction could be achieved by two stages, one a reduction of ten, the other of two. This translates to a 25 tap and a 17 tap filter. If an IIR is used, two biquadratic sections are simply cascaded.

Shown below are the data memory and multiplication requirements of the two approaches.

	FIR	IIR
Data memory (per channel)	43 words	6 words
Multiplications (per second)	1675	4000

Table 5.3 Number of channels handled dependent on filter type and memory configuration

	FIR	IIR
(a) No external memory (memory limited)	2	20
(b) External data (program space)	<40 (memory limited)	<50 (time limited)
(c) Peripheral RAM (see section 5.5.2)	<100 (time limited)	no benefit

The number of channels that could be implemented using the different approaches depends on the data memory and configuration of the TMS 32010 system.

Table 5.3 shows the tradeoffs.

- PROGRAMMING EQUATIONS

  V (K) = 2*[A/2*X (K) + (-B/2)*V (K-1) + (- C /2)*V(K - 2)]

  Y (K) = V (K) + 2*(D/2)*V (K 1) + V (K-2)

  V (K-2) ← V (K-1)

  V(K-1) ← V(K)

- DATA MEMORY ORGANIZATION

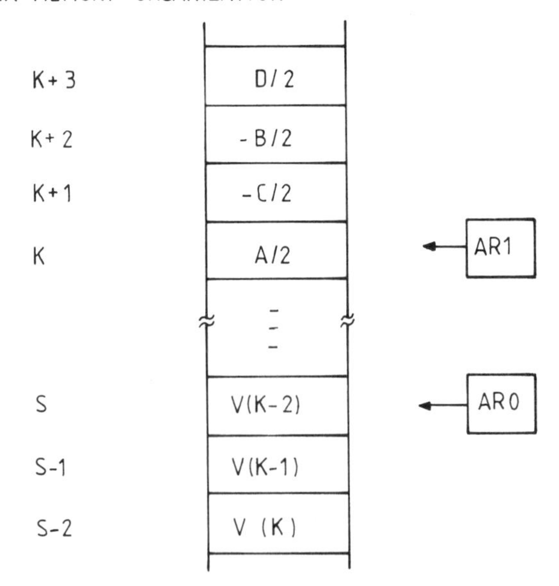

Figure 5.11 Memory organisation and programming equations for the biquad section of an IIR filter

5.6.2 Implementing the IIR filter

When used in a multiplexed mode it is more reasonable to design the filter to be a subroutine rather than a kernel program. To implement this and including scaling, the biquadratic filter will execute in less than 5 μs (the eleven instructions of the basic biquadratic filter take 2.2 μs).

Fig. 5.11 shows the programming equations and the data memory organisation to allow the IIR program 5.9 to execute. This program is shown with annotation and the status of the auxiliary registers at each instruction.

In summary, the IIR decimator using no external memory can handle over twenty channels of the DAS example requiring a modest

Before entry into subroutine:

ARP	AR0	AR1	Instruction	Comment
—	—	—	LT	T ← X(N), OR Y(N) FROM PREVIOUS STAGE
—	—	—	LAR 1 K	AR1 ← K
—	—	K	LAR 0 S	AR0 ← S
—	S	K	MAR 1	ARP ← 1
1	S	K	ZAC	ACC ← 0
1	S	K	CALL BIQUAD	

Biquad subroutine:

ARP	AR0	AR1	Instruction	Comment
1	S	K	MPY *+	P ← A/2*X(K)
1	S	K + 1	LTA −	T ← −C/2,ACC A/2*X(K)
0	S	K + 2	MPY *	P ← −C/2 * V(K − 2)
1	S − 1	K + 2	LTA *+	T ← −B/2, ACC ACC + P
0	S − 1	K + 3	MPY *−	P ← −B/2 * V(K − 1)
1	S − 2	,,	LTA *	T ← D/2, ACC V(K)/2
0	S − 2	,,	SACH *+,1	V(K) ← 2 * (ACC)
0	S − 1	,,	MAR *+	
0	S	,,	ZALH *−	ACC ← V(K − 2)
0	S − 1	,,	MPY *	P ← D/2 * V(K − 1)
0	S − 1	,,	LTD *−	ACC ← ACC + P, V(K − 2) ← V(K − 1)
0	S − 2	,,	LTD *+	ACC ← ACC + P, V(K − 1) ← V(K)
0	S − 1	,,	ADDH *	ACC ← ACC + V(K)
0	S − 2	,,	SACH *,1	Y(K) ← 2 * (ACC)
0	S − 2	,,	RET	

Program 5.9 TMS 32010 code for IIR digital filter

10 kHz A/D conversion rate (500 Hz/channel). The loading on the processor is about 15% and its uses all the data memory.

Using external program memory (for data space) up to fifty channels with a 25 kHz A/D conversion rate can be tackled. The processor would be fully loaded. The data table in the program memory would be approximately 300 words long.

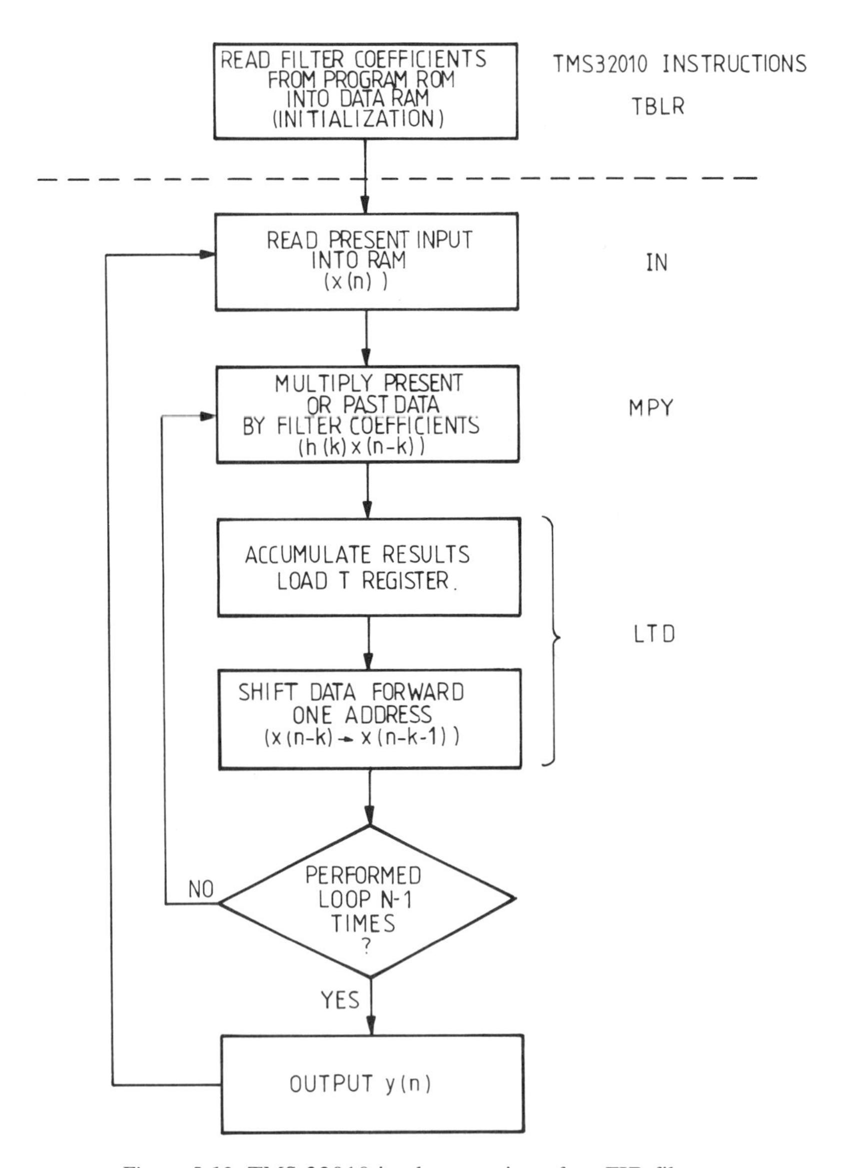

Figure 5.12 TMS 32010 implementation of an FIR filter

5.6.3 Implementing the FIR filter

The basic operations involved in an FIR construct are multiply, accumulate and shift. The two instructions of the TMS 32010(M) to perform this are LTD and MPY.

A block diagram of the implementation of an FIR filter on the TMS 32010(M) is shown in fig. 5.12.

The LTD instruction performs the equivalent of the LT, APAC and DMOV instructions in that order. The DMOV instruction performs a data move in the data memory to implement the discrete-time unit-delay operator. When using the LTD instruction, the order of the multiply and accumulate operations becomes important because data is moved while the calculation is taking place. The oldest discrete-time input variable must be multiplied by its associated constant in the difference equation and loaded into the accumulator first. Then the next more recent input (one unit-time less) is multiplied and accumulated until the entire difference equation is performed. The following example (program 5.10) implements a discrete-time FIR digital filter with the characteristic difference equation:

$$y(n) = A*X(n) + B*X(n-1) + C*X(n-2) + D*X(n-3) + E*X(n-4)$$

```
*
START ZAC        Initialise the ACC
      IN X0,PA0  Input data from port PA0 into data memory X0.
      LT X4      X(n-4)
      MPY E      E*X(n-4)
      LTD X3     ACC=E*X(n-4);
                 X(n-4)=X(n-3).
      MPY D      D*X(n-3)
      LTD X2     ACC=E*X(n-4)+D*X(n-3);
                 X(n-3)=X(n-2).
      MPY C      C*X(n-2)
      LTD X1     ACC=E*X(n-4)+D*X(n-3)+C*X(n-2);
                 X(n-2)=X(n-1)
      MPY B      B*X(n-1)
      LTD X0     ACC=E*X(n-4)+D*X(n-3)+C*X(n-2)+B*X(n-1);
                 X(n-1)=X(n).
      MPY A      A*X(n)
      APAC
      SACL Y     Store result in data memory Y.
      OUT Y,PA1  Output result to port PA1.
      B START
*
```

Program 5.10 FIR filter.

Note that $X0$, $X1$, $X2$, $X3$ and $X4$ are consecutive data memory locations in ascending order to take advantage of the LTD instruction.

This routine requires 18 cycles (3.6 μs) to execute.

5.7 Application to Spectral Analysis

One of the greatest strengths of the TMS 32010(M) is its performance in the area of spectral analysis using the fast fourier transform (FFT). There is a large market for portable spectrum analysers and currently many are based on analog bandpass filters giving a rough spectral estimate on some crude LED display for example. An application to engine monitoring and control is being researched by several major automobile manufacturers. The spectral signature of a family or 4 stroke engine has typically two spectral peaks in the 7 and 14 kHz regions. The effect of knock etc. within the cylinders causes shifting and reshaping of these peaks. A portable analyser based on the TMS 32010 for use by garages is now feasible.

Extrapolating the trends in cost and reliability, a serious possibility for a motor car model from 1987 onwards, would be an in-car analyser fixed as standard, providing, in addition to the spectral monitoring, the feedback control to optimise performance and engine life. In the telecom market the concept of the FFT providing a bank of matched tone filters has been used in TMS 32010 designs to provide tone detection in multiple tone FSK systems. The beauty of the FFT is its inherent ability to provide amplitude and phase information.

A simple block diagram showing a TMS 32010 based spectrum analyser is shown in fig. 5.13. The TMS 32010 in addition to the FFT has two functional blocks associated with it, windowing, and squaring and summing. Windowing is as much an art as a science. Although not mandatory, some form of window function is recommended to improve spectral resolution. The FFT is simply a time efficient method for implementing the DFT (discrete Fourier transform) and the inherent problems with the significantly large side lobes of the sin X/X function still have to be overcome. Windowing requires additional computational time, multiplying the time samples with the window function, but this can be tackled by the TMS 32010. The squaring and summing of the real and imaginary parts from the FFT is a relatively straightforward task for the device.

A variety of options for the system configuration of the TMS 32010 to perform the FFT are available. Using one TMS 32010, and only on-chip data memory, a 64 point FFT can be executed. 128 locations of the data RAM store the complex samples and the additional 16

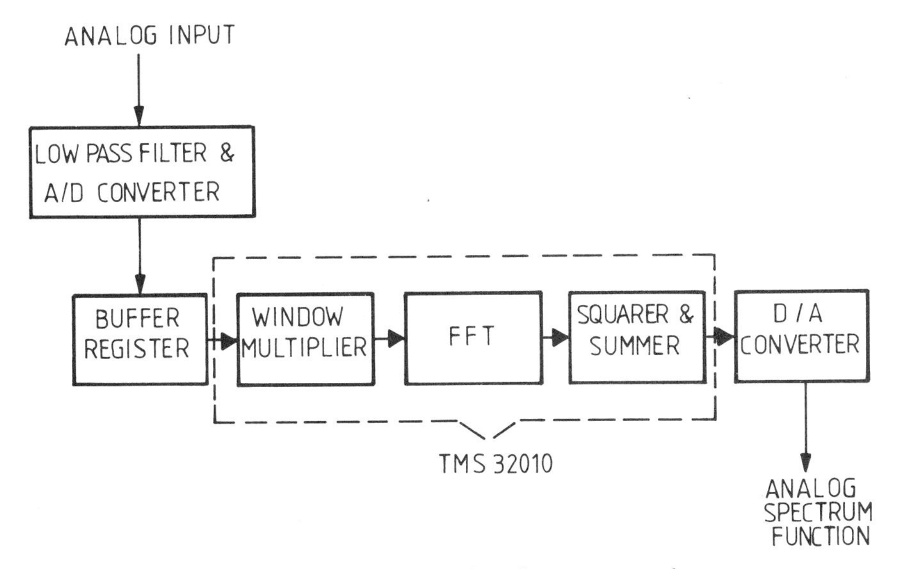

Figure 5.13 TMS 32010 based spectrum analyser

locations are for scratchpad. The speed of execution of the algorithm is dependent on the nature of the program code. If the code is predominantly looped as opposed to in-line the benefit of needing only on-chip program ROM (i.e. no memory expansion) is balanced by a reduction in execution speed. By using straightline code, and the majority of the 4K word address space, a factor of 2 reduction in execution time is typical. Table 5.4 gives the execution time for a straightlined radix 4 64 point complex FFT as 580 μs.

The ability to synchronise two TMS 32010s in a multiprocessing configuration is well demonstrated and utilised in providing higher performance spectrum analysers. Fig. 5.13 shows the system layout. Both services are programmed with identical radix 2 64 pt FFT looped code (no off-chip memory used) and operate in a 'ping pong' fashion. Whilst one TMS 32010 carries out the FFT operation on data K − 1, the second TMS 32010 performs both the windowing calculations on the current data K and also performs the squaring and summing of data K − 2. Once the FFT (most time consuming) has finished the two swap functions and so on. The external logic controlling this action is simple. Because of the configuration the data buses can also be wired together. D/A write and A/D read signals are provided. Although not totally code optimised an experimental system provided 1.5 ms/FFT, giving a sampling period of 1.5/64 ms translating to a sampling frequency of 43 kHz. Given Nyquist conditions the maximum input

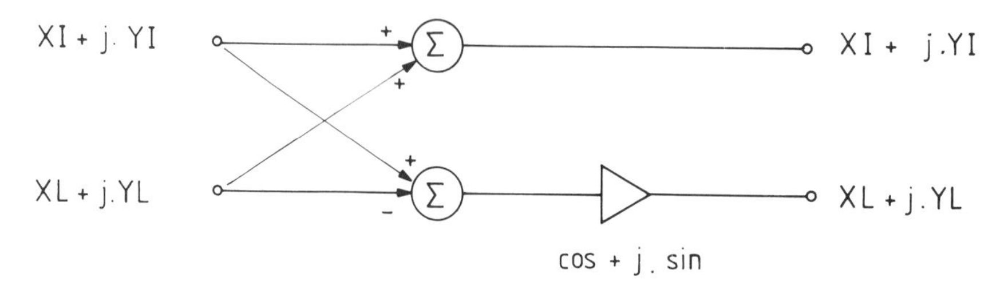

Figure 5.14 FFT butterfly

frequency is 21.5 kHz. By using the full memory expansion option on both devices performing straightlined code, a reduction in execution speed of more than one half, to 737 μs/FFT, has been achieved. This gives an input bandwidth of approximately 43 kHz.

FFTs beyond 64 point are possible using the TMS 32010 by multiprocessing with two or more devices. It is feasible to parallel 64 TMS 32010s, each performing a 32 point transform for a 1024 point FFT. This would allow straightline coded TMS 32010s to be used, with no off-chip program memory, and give a 1024 point computation time equal to a 32 point time, less than 400 μs.

A method of implementing larger point FFTs on one TMS 32010 is to use the extended data memory addressing technique described in section 5.5.2. By storing the 1024 complex (2048) real) points in off-chip data RAM, input/output instructions can bring data into on-chip RAM for computation. The FFT butterfly shown in fig. 5.14 and program 5.11 provides for a 'scrambled output' FFT. 'Scrambled output' more aptly describes the decimation in frequency algorithm. The time samples are taken in order and the output frequency samples are scrambled. The decimation in time (scrambled input) style of algorithm can equally be implemented. These scrambled Input/Output approaches provide an 'in place' capability, i.e. samples are taken from memory, processed and returned back to the same locations. A non-scrambled approach, although providing higher speed, does not give an 'in place' algorithm and as such requires larger memory spaces. The butterfly shown also provides a scaling function. Values often appear as elements to be summed in the FFT. This could cause overflow. It is recommended that some form of scaling is used at each stage to avoid this, e.g. by half. The shifter on the TMS 32010 is ideal for this function. Although inherently a left shifter the requisite data right shift is performed by a left shift of 15.

Table look-up time can be reduced by performing all butterflies for a given complex weight. Table 5.4 shows the execution time for such a

```
                            LAC     XI,15
                            ADD     XL,15
OUT    I,PA0                SACH    XI
IN     XI,PA1               SUBH    XL
IN     YI,PA1               SACH    XT,1
OUT    L,PA0                LAC     YI,15
IN     XL,PA1               ADD     YL,15
IN     YL,PA1               SACH    YI
                            SUBH    YL
                            SACH    YL,1
                            LT      XT
       (BUTTERFLY)          MPY     COS
                            PAC
                            LT      YL
                            MPY     SIN
OUT    I,PA0                SPAC
OUT    XI,PA1               SACH    XL
OUT    YI,PA1               MPY     COS
OUT    L,PA0                PAC
OUT    XL,PA1               LT      XT
OUT    YL,PA1               MPY     SIN
                            APAC
                            SACH    YL
```

Program 5.11 FFT RADIX 2 butterfly, with scaling using peripheral data RAM

Table 5.4 FFT execution times

Number of Complex Points	FFT Time
64	580 μs
256	6.6 ms
512	11.5 ms
1024	33 ms

single TMS 32010 1024 point FFT system is 33 ms. The execution time is proportional to $N \operatorname{Log}_2 N$ (N is the number of points). Doubling N leaves $\operatorname{Log}_2 N$ reasonably unaffected giving execution time approximately proportional to N. One could extrapolate to give a 2048 point FFT in 66 ms.

5.8 TMS 320 Family Direction

The TMS 320 family is developing with major activity in five areas, current device expansion, support, peripherals, modules and future generation devices.

The initial releases of the TMS 32010(M) were available in ceramic and plastic 40 pin DIL packages. Due to the very large military and certain industrial markets the devices are being specified to MIL 883B process flow and −55°C to +125°C temperature range. The bar size is compatible with current ceramic chip carrier packages and this option will also be available. An industrial −40°C to +85°C range option is another variant. The NMOS devices have a typical power dissipation of less than 1 W. However, for applications requiring a lower power rating, other technologies, e.g. CMOS, have been evaluated for other TMS 32010(M) variants. Texas Instruments has started to release information on 'off the shelf' customised ROM variants of the TMS 320M10. Many design teams will not want or perhaps will not be able to design filter or FFT routines for example. Having on-chip modules that can be used by off-chip code is rapidly becoming an attractive option.

Although the TMS 32010(M) will be superseded as the TMS 320 family's most powerful DSP processor, it is reasonable to expect it to have a long market life. As the cost falls, and the engineering expertise rises, more application areas will open to the TMS 32010(M). It could well become another TMS 1000, the original calculator chip from a decade ago, which still sells in prodigious volumes into games, washing machines, etc.

Section 5.4 described current development support. Several manuals and books discussing specific design problems are being written. During 1983 Texas Instruments started a university programme supporting key establishments in their research and undergraduate teaching. One of the first design utilities (filters) has come from initial links with Rice University, Houston, Texas. Several U.S. and European third parties, including universities, have begun to release software support products on to the market. This activity will be encouraged for the current and future family members.

Development of a series of modules for DSP applications based on the TMS 320M10 has been started. For example a board with a standard bus interface with the requisite code for voice store and forward and also recognition was made in 1983. A real time vocoder based on LPC is another attractive product for those people who need the speech functionality and could not develop their own algorithm. It is expected that third party modules will become available performing such tasks as spectral analysis or attached processing.

Texas Instruments has two devices due for release in 1984 to perform the analog input and analog output function. The devices are the first intelligent analog peripherals providing A/D and D/A converters, associated filters and control logic. The devices are I/O bus compatible with the TMS 32010(M).

The analog input channel (AIC) using switched capacitor technology has a programmable antialiasing low pass filter. The 16-bit two's complement A to D converter can be used in a free running or TMS 32010(M) triggered mode. Two analog channels can be connected with an on-chip multiplexer routing to the A to D converter. The antialiasing filter is very comprehensive, comprising three individual filters. The first low pass filter conditions the analog signal before it enters the discrete-time switched capacitor filter. The third filter removes the spectral components above the Nyquist frequency of the discrete time filter.

The analog output channel (AOC) has a FIFO buffer for the digital to analog (DAC) converter. A $\sin(x)/x$ compensation filter following DAC is provided for sampling attenuation offset. A smoothing output filter with programmable bandwidth is on chip. Two output ports (mirroring the AIC) are driven via an output multiplexer.

Both parts have been designed to adequately cope with the processing bandwidth of the TMS 32010(M) but also for the next generation devices.

In order to bring out the next CPU in the family in a shorter timescale than sequential definition and MOS design time, redundancy was built into the definition of the TMS 32010(M).

The second generation device in the family will be software upward compatible with the TMS 32010(M). The instruction set will be significantly enhanced but it is possible to take existing TMS 32010(M) source code and site it on the next generation part without re-coding. By close analysis of the TMS 32010(M) register redundancy, the ability to significantly increase both data and program memory address spaces can be seen. Larger on-chip data RAM will be an added benefit. The architecture supports up to eight auxiliary registers. Increasing from the current two, AR0 and AR1, on the TMS 32010(M) will allow full indexed addressing. A single cycle multiply/accumulate along with floating point support will be given. The device will be several times faster than the TMS 32010 due to cycle time reduction and instruction set enhancements and architectural utilisation.

The single accumulator definition of the TMS 320 architecture has provided an excellent start for DSP microcomputers. Looking beyond 1985, new architectures will be adapted to adequately tackle the demands of, for example, digital hifi or TV. Work on devices to address these and other potential DSP markets is under way.

The TMS 320 family has the signs of becoming as synonymous with DSP microcomputers as the 8085 has with general purpose microprocessors. In its short life it has already gained a large market acceptance on both technical and support grounds.

CHAPTER 6
SUMMARY AND FUTURE TRENDS

D. J. Quarmby, Loughborough University

6.1 Device Selection

We have provided a great deal of information which should aid the reader in selecting a signal processing chip for any particular application. Invariably, such a choice involves many factors other than purely technical ones. The availability of support equipment and expertise within an organisation is probably the most important factor. We could summarise the technical considerations under two main headings; system size and cost, and design costs.

6.1.1 Chip Count

In spite of its age, the Intel 2920 can provide the lowest chip count in very small systems. It is highly self-contained, and can be viewed as an analog component whose function is controlled in a very flexible manner by its instruction EPROM. It is best suited to these small systems, performing perhaps a filtering operation on one or more signal channels. As with any other analog component, a number of devices can be linked in a system, but if this number exceeds two or three it is likely that the use of a single, more powerful device, will lead to a system which has a lower chip count and better performance.

The NEC 77P20 can provide very neat total systems involving around ten devices, including a small host computer based on a standard microprocessor such as Intel's 8085 or 8088. Analog interfaces in such a system could be attached either to parallel ports on the host computer, or directly to the serial ports of the signal processor. Additional functions which can be provided by a small host computer are the generation of timing signals – particularly to control the sampling interval – and the buffering of input/output in a larger

RAM than is available on the signal processor. Low chip counts are retained in systems involving a number of 77P20s. A typical example is the channel vocoder reported by Feldman of the MIT Lincoln Laboratory (ref. 6.1) involving six 77P20s and an 8085 host. The total chip count is 26 integrated circuits, and the board dissipates 11.5 watts. The LPC vocoder designed in the same laboratory (ref. 6.2) is a very similar system organisation, but uses only three 77P20s, with a total chip count of 16.

The chip count in prototype TMS 320 systems is somewhat higher, since initial work will involve a program in off-chip memory. The step into a mask-programmed device is a major one which will never be taken by the majority of designers. The use of standard EPROMs for the program memory would slow the 320 down, and so fast static RAM is the usual choice for prototyping. This RAM has to be loaded by a host processor, and additional devices are needed for switching the address source. A good example of a small system is the evaluation board from Texas Instruments, which together with the analog I/O board forms a powerful general-purpose system. The design of a general-purpose digital speech processing system has been reported by Daly and Bergeron (ref. 6.3). This system involves memory shared by a TMS 32010 and an Intel 8089 input/output processor. No chip count is given, but the block schematic would indicate a rather large single board system.

No meaningful and lasting hardware cost assessment can be made, as the system costs are currently dominated by the high cost of signal processor chips. These are rapidly being reduced.

6.1.2 Design Effort

The cost of hardware design is closely related to chip count. Software design for signal processor chips is usually a substantial proportion of total project cost.

When using the 2920 there are clear limits to the software cost imposed by the 192 instruction limit and the non-existence of branching instructions. The process of building coefficient values into the code is lengthy if done manually, but filter design has been greatly simplified with software to automatically generate the code from a filter specification.

Multiple-field, 'horizontal' instructions present the programmer with an opportunity to create surprisingly fast programs. This is particularly true of the OP instruction format of the NEC 7720. This detailed level of programming requires a good deal of skill to be properly exploited. Certainly, any programmer who does not have experience in writing assembler language for a more conventional computer would find it difficult to write for the 7720. Code generation from higher levels has been touched upon in chapter 4, but at present this is not the approach which most systems designers would take. The usual reasons for writing at the assembler level apply even more strongly to real-time signal processing. Speed of execution is paramount, and a close knowledge of timing details is important where hardware interactions are taking place. Software debug time can be lengthy, but the manufacturers provide excellent in-circuit emulators which minimise this problem. A combination of single-step testing and the use of breakpoints is the first line of defence. A second tool which is often very useful is the D/A converter used to output values of intermediate variables. It is often worth adding D/A converters to a system purely for assistance in debugging programs.

In using the NEC 7720 as a slave processor, there is usually close co-operation with a host processor to be considered. The level of interaction between the host and the TMS 320 is likely to be lower, since the 320 acts as a master processor. Usually the interaction would take place via shared RAM, and might involve the host only in loading a program to this memory.

The TMS 320 has a relatively large program memory, and the more conventional 'vertical' style of instruction which makes the programming quite similar to the use of a normal assembler language. Although the need for execution speed and close control still apply, it is more feasible to consider the use of high level programming. The high level design techniques, used by L.R. Morris and referred to in chapter 2, go some way along this route, and for the TMS 320 it may well be that a compiler could be written for a conventional high level real-time language.

6.2 Towards Floating-point Architectures

The benefits of carrying out signal processing arithmetic in floating point are great. It is the 'natural' representation for variables covering a wide dynamic range. It simplifies the task of the programmer, opening the work to a wider group of people. Programs are likely to be more reliable in that the risk of calculation overflow is substantially reduced. Manufacturers have recognised these benefits, and devices to perform floating-point arithmetic at similar speeds to the fixed-point devices which we have described are currently being announced. At present, these arithmetic units have both fixed-point and floating-point functions. Fixed-point processing, and particularly integer processing, are still needed in control (e.g. loop counting) and in addressing. It would be an interesting development to see separate arithmetic units provided for these quite different uses of arithmetic, as happens in some of the large array processors.

The most sophisticated floating-point devices are intended for micro-coded multiple-chip signal processors, and a particularly useful range of devices is currently being made available by TRW. Recent papers from staff members of TRW describe the range of currently three chips (refs. 6.4, 6.5, 6.6). TRW have adopted a 22-bit floating-point format, using 16 bits in the mantissa and 6 bits in the exponent. This exponent provides over 380 dB of dynamic range which comfortably encompasses signal processing needs.

The first of the three chips is the floating point adder, the TDC 1022. It can add, subtract and convert between 16-bit fixed-point and 22-bit normalised floating-point formats. Its function would normally be controlled by micro-instructions which can be sequenced at a 10 MHz clock rate.

The second chip, TDC 1042, is a multiplier which supports both the 22-bit floating-point format and 16-bit fixed point, with a 32-bit product available (fixed-point or mantissa of floating-point). This device can also be clocked at 10 MHz.

The third arithmetic unit is the TDC 1033, and it includes sixteen 22-bit registers. Where the TDC 1042 and TDC 1022 can often be used in a stand-alone situation, the TDC 1033 is clearly intended for use in a microcoded system. Its instruction set includes the instructions of the AMD 2901 and 2903 bit-slice fixed-point ALUs. Thus it provides a wide variety of fixed-point arithmetic as well as the floating-point addition and subtraction which are the main innovations. The floating-point facilities and instruction formats, are those of the TDC 1022. The last of the four arithmetic functions, division, can be carried out in the 22-bit floating-point format in a sequence of 15 cycles of the 6 MHz clock. This compares favourably with the times for fixed-point division on the current signal processing chips.

The paper by Winter and Yamashita (ref. 6.6) describes a single board system incorporating all three of these chips, together with a 'flash' A/D converter and a fast D/A converter. The function of the board is to provide two FIR filter stages, one of 32 taps and one of 14 taps, on a signal bandwidth of 100 kHz, maintaining the full dynamic range of the A/D converter – extended from its basic 9 bits by use of sample accumulation.

Systems involving these TRW devices will be relatively large, comparable with the bit-slice systems discussed in chapter 1. A single-chip signal processor with floating-point arithmetic has been described by workers at the Hitachi Central Research Laboratory (ref. 6.7). This device includes a floating-point multiplier, whose output is pipelined into a floating-point/fixed-point ALU. Both units use a 4-bit exponent when dealing with floating-point numbers, the multiplier working on a 12-bit mantissa, and the ALU on a 16-bit mantissa. In many other respects this device resembles the NEC 7720. It has a 512 × 22-bit instruction store on chip. Separate data RAM (200 16-bit words) and data ROM (128 16-bit words) are also provided on chip. The instruction cycle is 4 MHz, and two 20-bit accumulators are associated with the floating-point ALU. Interfaces are 16-bit parallel, and serial I/O lines. Quoted execution times are 1.5 μs for a biquad filter, 8.76 μs for a 32-tap transversal filter. CMOS circuitry is used to keep power consumption low (typically 200 mW), and the package size is 40 pins.

6.3 Towards Parallel Processing Architectures

Multiple chip systems using the present generation of signal processors (and indeed of most processors) rely on a loose coupling of the chips. This means that one chip will perform one job which can be cleanly separated from the rest of the system. The pipeline arrangement is a typical one, in which each chip performs its own job on the current piece of data, while the chip following it in the chain is performing its job on the previous piece of data, etc. It is rare that a signal processing pipeline can exceed four or five stages, unlike Henry Ford's factory pipelines where many stages are used to go from component to finished motor car. One reason for limiting the pipeline length is the delay which such an arrangement involves, which can become prohibitive in some real-time processing, such as that of conversational speech. Delays are avoidable in situations where processing can be performed in parallel. An example in which this can be done is the filter bank, where a number of chips can be used, each providing one filter output from an input which is common to all. A similar situation arises in pattern comparison, where one unknown is to be compared with a large number of stored patterns. The arithmetic of the comparison (usually a distance measurement) can be done in parallel, with one chip assigned to a subset of the total set of stored patterns.

When higher levels of interaction between signal processors are required, special hardware facilities for fast interchange of data are particularly useful. A British-designed chip, the CRISP signal processor from STL, provides such facilities. This device, like the TMS 320, uses external memory for its program store, accessed through the usual address and data lines. However, a second data bus, the 'system' bus, is provided to facilitate very fast data transfer from the internal RAM of one chip to the internal RAM of another. Final specification and production of this device is scheduled for 1984.

Still further in the future is another recently announced British innovation. This is the IMS T424 transputer from INMOS. Provision for parallel processing has been one of the fundamental design objectives of INMOS. Their approach is an interesting one, and very fundamental. They have first standardised on a programming language, 'occam', in which the 'process' is the main unit of computation. Each process has its own program and data, but can communicate via input and output channels with other processes. In theory, processes can proceed in parallel – provided that adequate hardware is used. Input and output to the outside world is treated in the same way as transfers between processes. Occam is a neat, high level, well-structured language, and INMOS intend that it will be the lowest level at which the user will program their devices. The T424 transputer is intended to efficiently implement the occam language. It will have 32-bit arithmetic addressing external memory via a 32-bit multiplexed address/data bus. Four K bytes of internal fast RAM will be provided, and four serial links to other transputers will be available to provide the inter-process communication when processes are implemented on different chips. Peripheral input/output is to be via an 8-bit bus, to ease interfacing with existing devices.

The transputer is not intended primarily as a signal processing device, but there is little doubt that this will be one of the first application areas for it.

References

6.1 Feldman, J.A. (1982) 'A compact digital channel vocoder using commercial devices' *Proceedings of IEEE ICASSP '82, Paris*, 1960–1963.

6.2 Feldman, J.A., Hofstetter, E.M. and Malpass, M.L. (1983) 'A compact, flexible LPC vocoder based on a commercial signal processing microcomputer' *IEEE Trans ASSP-31*. No. 1, 252–257.

6.3 Daly, D.F. and Bergeron, L.E. (1983) 'A programmable voice digitiser using the T.I. TMS-320 microcomputer', *Proceedings of IEEE ICASSP '83, Boston*, 475–478.

6.4 Schirm IV, L. (1981) 'A family of high speed, floating-point arithmetic chips' *Proceedings of IEEE ICASSP '81, Atlanta*, 374–377.

6.5 Eldon, J. (1983) 'A 22-bit floating-point registered arithmetic logic unit' *Proceedings of IEEE ICASSP '83, Boston*, 943–946.

6.6 Winter, G.E. and Yamashita, R.R. (1983) 'A single board floating-point signal processor' *Proceedings of IEEE ICASSP '83, Boston*, 947–950.

6.7 Hagiwara *et al.* (1982) 'A high performance signal processor for speech synthesis and analysis' *Proceedings of IEEE ICASSP '82, Paris*, supplement.

INDEX

NOW . . . Announcing these other fine books from Prentice-Hall—

16-BIT MICROPROCESSORS, by Ian Whitworth. Here is a highly readable survey of 16-bit microprocessors including system software and multiple processor systems. In addition to telling how 16-bit microprocessors evolved from 4-bit and 8-bit chips, it also explains control systems and signal processing applications.

$15.95 paperback

DIGITAL IMAGE PROCESSING, by Gregory Baxes. With this book as a guide, anyone can learn how to perform image processing—even those without an engineering background or knowledge of calculus. Specific yet nontechnical, it covers a wide range of capabilities and applications. Fully illustrated.

$14.95 paperback $22.95 hardcover

To order these books, just complete the convenient order form below and mail to **Prentice-Hall, Inc., General Publishing Division, Attn. Addison Tredd, Englewood Cliffs, N.J. 07632**

Title	Author	Price*

Subtotal ____________

Sales Tax (where applicable) ____________

Postage & Handling (75¢/book) ____________

Total $ ____________

Please send me the books listed above. Enclosed is my check ☐ Money order ☐ or, charge my VISA ☐ MasterCard ☐ Account # ____________

Credit card expiration date ____________

Name ____________

Address ____________

City ____________ State ____________ Zip ____________

**Prices subject to change without notice. Please allow 4 weeks for delivery.*